AF824098

SIMULATION IN
ELECTRICAL POWER SYSTEM

SIMULATION IN
ELECTRICAL POWER SYSTEM

Dr. Ashok Kumar Tiwari
General Manager
Electricity Distribution Company

NAMYA PRESS.com

Published by

Namya Press

India: 213, Vardan House, 7/28 Ansari Road Darya Ganj - Delhi India 110002
USA: 10685-B Hazelhurst Dr. # 31013 Houston, Texas USA - 77043
Email: namyapress@gmail.com
Website: https://namyapress.com

Edition	:	First Published in 2023 (Namya Press)
Title	:	Simulation in Electrical Power System
Authors	:	Dr. Ashok Kumar Tiwari
ISBN	:	978-93-90124-99-2
Copyright	:	© Dr. Ashok Kumar Tiwari. 2023 All Rights Reserved

Dedicated to my

Two Gurus

Dr. M. P. Chourasia in Power System Engineering

Dr. N. TataRao in Power System Management

"Even if the Earth were turned into paper, all forests into pens, and all oceans into ink, the virtues of a Guru cannot be fully written. My infinite respects to both of you. Although you have both shed your physical bodies, I always feel your presence near me."

PREFACE

Simulation has become an indispensable tool in the field of electrical engineering. With the rapid advancement of technology and increasing complexity of electrical systems, simulation has proven to be an effective method for analysing and designing electrical systems.

This book aims to provide a comprehensive guide to simulation in electrical engineering. It covers a wide range of topics, from basic concepts and principles of simulation to advanced techniques and applications. This book is intended for students, researchers, and professionals in the field of electrical engineering who are interested in simulation as a tool for analysing and designing electrical power systems.

Book begins with an introduction to simulation and its applications in electrical engineering. It then covers the different types of simulation methods and techniques, including analytical, numerical, and symbolic simulation. It also covers simulation tools and software commonly used in electrical engineering, such as MATLAB-Simulink, Mipower, CYME and ETAP, with prime focus on MATLAB.

This book then delves into specific applications of simulation in electrical engineering, such as power systems. It includes four practical examples and case studies in power system to illustrate the concepts and techniques presented in the book.

Finally, this book discusses the challenges and limitations of simulation in electrical engineering, as well as emerging trends and future directions. It provides insights into the current state of the art in simulation and its potential for future advancements in the field of electrical engineering .

Overall, this book aims to provide a comprehensive and practical guide to simulation in electrical engineering, covering both fundamental principles and advanced techniques. It is hoped that this manuscript will serve as a valuable resource for students, researchers, and professionals in the field of electrical engineering.

I would like to express my sincere gratitude to the following individuals who provided their assistance, guidance, and support in the development of this book:

Firstly, I would like to thank the Supreme Being and my family Prof. Bharti Sharma, Tripti, Ritwik, Priyadarshita and Shreyansh Shekhar for their constant support and encouragement throughout the writing process. Their unwavering belief in me and my abilities has been a great source of inspiration.

I am also deeply grateful to my mentors, colleagues and peers in the field of administration, academics and engineering, who generously shared their insights and expertise with me, Specially Shri Pankaj Agrawal IAS, Prof. Sunaina Singh (Hon.Vice Chancellor, Nalanda University), Dr. C P Gupta (Professor, IIT Roorkee), Dr. Sitaram Soni (Professor, Flame University Pune), Dr. Santosh Vishwakarma (Professor IIT Indore), Shri Manjeet Singh MD MPPGCL, Shri Rajeev Gupta CGM MPPMCL, Dr. Vivek Chandra GM, Dr. G S Khanooja GM MPPMCL, Shri Praveen Gargava ACE MPPTCL and Dr. Himanshu Shrivastava E.E. MPPTCL.

Special Thanks goes to Dr. A K Sharma, Dr. Anil Kori, Dr. Shailja Shukla, Dr. Ranjana Singh, Dr. A Pachauri, Dr. Priti Jain, Dr. Bhavna Jhariya, Dr. Basant Chaurasiya and Prof. Nitin Saxena from my alma mater Jabalpur Engineering College Jabalpur, Dr. I K Khanna from Takshshila Institute of Engineering and Technology Jabalpur, Dr. Manish Choubey, Dr. Rajeev Kumar Chouhan, Prof Vivek Anand and Prof Nidhi Mishra from Gyan Ganga Institute of Technology and Science Jabalpur. Their feedback and suggestions were invaluable in shaping the content and direction of this book.

The author would like to express distinct gratitude to Dr. Rahul Banerjee, Director LNM Institute of Information Technology Jaipur, Dr. Rakesh Singhai RGPV UIT Shivpuri, Prof. Rama Shukla of AKS University Satna, Er Diwakar Singh Hon. Chancellor and Dr. Biplab Paul Hon.Vice Chancellor Sardar Patel University Balaghat, Dr. A K Mishra Hon.Vice Chancellor and Dr. Vinita Saluja Hon. Pro VC Mangalayatan University Jabalpur and Dr. R C Mishra Hon. Vice Chancellor Mahakaushal University Jabalpur for their cooperation and well wishes.

Thanks to my special friends Shri Anupam Maheshwari, Dr. Anurag Trivedi Principal IGEC-Sagar, Dr. Sanjeev Verma, Principal Scientist CIRC Meerut, Dr. Rudra Rameshwar Professor Thapar Institute of Engineering and Technology Patiala, Dr. T K Tripathi, V K Sharma and Dr. Shishir Dutta of Coal India Ltd. for their continuous encouragement.

I always learned from my valued students a lot, thanks to Abhishek, Shikha, Shweta, Parul, Kirti, Akash, Navita, Meghna, Abhijaat, Diksha, Dipti, Akshat, Neha, Akriti, Ayush and Nikunj. Thanks to Abhilasha and Sanjana for typing the manuscript and Jasvinder for Drawing figures.

I would like to extend my appreciation to Shri Rajesh Sharma and the staff at the publishing company Namya Press for their support and guidance in bringing this book to fruition. Their professionalism and dedication to producing a high-quality product were instrumental in the success of this project.

Finally, I would like to express my heartfelt thanks to the readers of this book. It is my hope that this book will be a valuable resource to all those who seek to enhance their knowledge of simulation in electrical engineering. Thank you for your interest and support.

Dr. ASHOK KUMAR TIWARI

CONTENTS

CHAPTER - 2 : POWER QUALITY

1

SIMULATION IN ELECTRICAL ENGINEERING

1.1 What is Modelling in Engineering ?

In engineering, modelling refers to the process of creating a simplified representation of a system or process. The goal of modelling is to gain a deeper understanding of the behaviour of the system or process, and to use this understanding to design and optimize the system or process.

Modelling in engineering can take many forms, depending on the type of system or process being studied. Some common types of models used in engineering include:

1. **Mathematical models:** Mathematical models use mathematical equations to describe the behaviour of a system or process. These models can be used to simulate the behaviour of the system or process under different conditions, and to optimize the performance of the system or process.
2. **Physical models:** Physical models use physical components to represent the behaviour of a system or process. These models can be used to test the behaviour of the system or process under different conditions, and to optimize the design of the system or process.
3. **Computer models:** Computer models use software programs to simulate the behaviour of a system or process. These models can be used to test the behaviour of the system or process under different conditions, and to optimize the performance of the system or process.

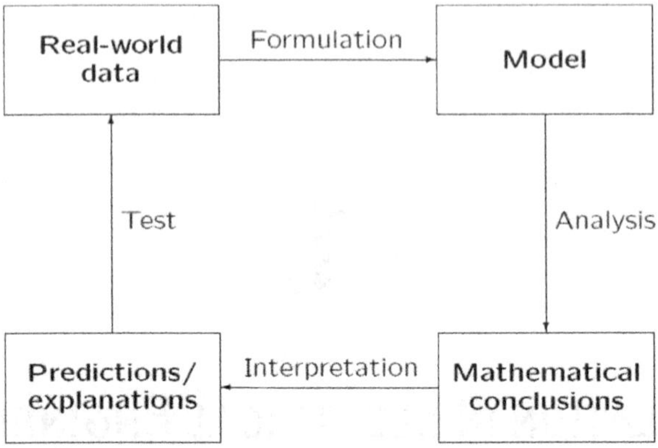

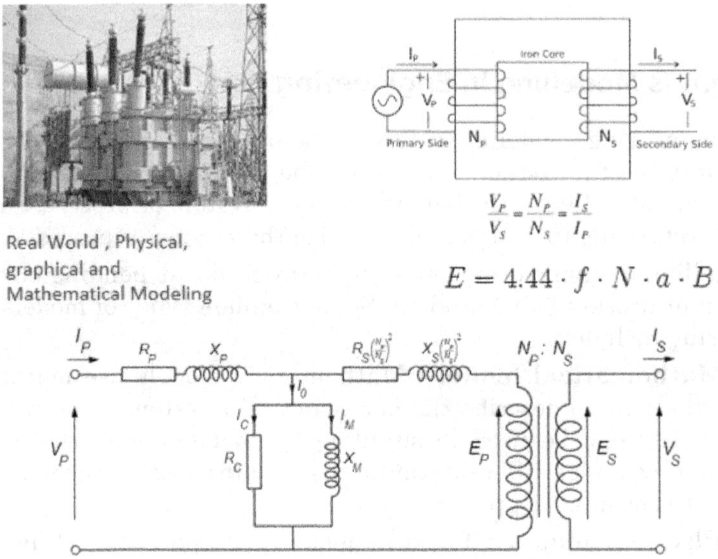

Overall, modelling is an important tool in engineering, allowing engineers to gain a deeper understanding of the behaviour of complex systems and processes, and to design and optimize these systems and processes to meet specific performance requirements.

1.2 What is Simulation?

Simulation is the process of creating a computer model of a real-world system or process and using it to predict or observe its behaviour under different conditions. The computer model can be used to replicate the functioning of a real-world system, including its inputs, outputs, and variables, and the interactions between them. Simulation can be used in many fields, including engineering, science, economics, and social sciences, to explore and analyze different scenarios, test hypotheses, and optimize performance.

Simulation models are often based on mathematical equations that describe the behaviour of the system being modeled. These equations may take into account factors such as physical laws, statistical distributions, or other variables that affect the system's behaviour. Once the model is created, it can be used to simulate the system's behaviour under different conditions, such as changing input values, adding or removing variables, or altering the parameters of the model.

Simulation has many benefits, including the ability to test and refine designs, optimize performance, and evaluate potential risks or uncertainties. It can also be used to explore hypothetical scenarios, such as the impact of changes in policy, technology, or environmental conditions. Simulation is often used in conjunction with other methods, such as experimentation, to gain a deeper understanding of complex systems and processes.

1.3 Simulation in Engineering

Simulation is a powerful tool used in engineering to model complex systems and analyze their behaviour under different conditions. It involves creating a computer model of the system being studied and then using the model to simulate its behaviour under various scenarios. The goal of simulation is to gain a better understanding of the system and to optimize its performance.

In engineering, simulation is used in a wide range of applications. For example, it can be used to simulate the behaviour of a car in a crash, to predict the performance of a new aircraft design, or to analyze the flow of fluids in a chemical process. Simulation allows engineers to test and refine their designs in a safe and controlled environment, without the need for costly physical prototypes.

Simulation software has become increasingly sophisticated in recent years, allowing engineers to model complex systems with a high degree of accuracy. It is now possible to simulate the behaviour of a system under a wide range of conditions, including extreme weather events, seismic activity, and other factors that would be difficult or impossible to reproduce in the real world.

One of the key benefits of simulation is that it allows engineers to explore different design options and test their performance under different conditions. This can help to identify potential problems early on in the design process, and to optimize the system for maximum performance and efficiency.

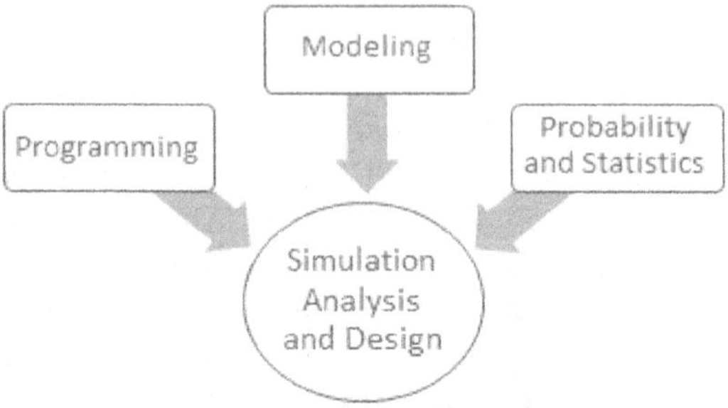

In conclusion, simulation is a valuable tool in engineering that allows engineers to model and analyze complex systems in a safe and controlled environment. It has become an essential part of the design process, allowing engineers to test and refine their designs before they are implemented in the real world. As simulation software continues to improve, it is likely that it will become an even more important tool in the field of engineering.

Example: Driving Simulator

Analytical, Numerical and Symbolic Simulation

Analytical simulation is a mathematical method that uses equations to model and simulate a system's behavior. It involves the use of mathematical formulas to derive solutions that can predict a system's future behavior. Analytical simulations are often used in physics, engineering, and other scientific fields where precise calculations are required. The main advantage of analytical simulation is its ability to provide accurate results in a short amount of time. However, it is limited to systems that can be represented mathematically.

Numerical simulation involves the use of computer algorithms to solve complex mathematical equations. It involves breaking down the system into small parts and using numerical methods to simulate each part's behavior. Numerical simulations are often used in fields such as physics, engineering, and computer science. The advantage of numerical simulation is its ability to simulate complex systems that cannot be represented analytically. However, it requires a significant amount of computing power and can be time-consuming.

Symbolic simulation involves the use of computer programs that can manipulate mathematical expressions and equations. It involves the use of symbolic reasoning to simulate a system's behavior. Symbolic simulation is often used in fields such as mathematics, computer science, and artificial intelligence. The advantage of symbolic simulation is its ability to provide exact solutions to complex problems. However, it can be computationally expensive and is limited to systems that can be represented mathematically.

1.4 Differences between Modelling and Simulation

Modelling and simulation are two related but distinct concepts in engineering. Here are five main differences between Modelling and simulation:

1. **Purpose:** The purpose of Modelling is to create a simplified representation of a system or process in order to gain a deeper understanding of its behaviour. The purpose of simulation is to use this model to predict the behaviour of the system or process under different conditions.
2. **Scope:** Modelling is typically a more abstract and theoretical process that focuses on creating a conceptual representation of a system or process. Simulation, on the other hand, is a more practical and applied process that involves running the model to generate predictions of real-world behaviour.
3. **Inputs:** In Modelling, the inputs are theoretical or conceptual variables that are used to build the model. In simulation, the inputs are actual data or measurements of the system or process being modelled.
4. **Output:** The output of Modelling is a conceptual representation of the system or process that can be used to gain insight and understanding. The output of simulation is a set of predictions or results that can be used to evaluate the behaviour of the system or process under different conditions.

5. **Iteration:** Modelling is typically an iterative process that involves refining the model based on new data or insights. Simulation, on the other hand, is typically a one-time process that generates a set of results based on a given set of inputs.

Overall, Modelling and simulation are complementary tools in engineering that are used to gain insight and understanding into complex systems and processes. While they share some similarities, they have distinct differences in terms of purpose, scope, inputs, output, and iteration.

1.5 Simulation in Power System

Simulation is an important tool used in power system engineering to model and analyze the behaviour of complex electrical networks and systems. Power system simulation involves creating a computer model of the power system, which includes generators, transformers, transmission lines, and other components, and using the model to simulate the system's behaviour under different operating conditions.

Power system simulation can be used to analyze and optimize the performance of the power system, including its stability, reliability, and efficiency. It can also be used to test and refine new designs, evaluate the impact of changes in operating conditions or system parameters, and predict the behaviour of the system under different scenarios, such as faults or disturbances.

One important application of power system simulation is in the design and analysis of control systems. Control systems are used to regulate the flow of power through the system, ensuring that it remains stable and reliable. Simulation can be used to model the behaviour of different control systems and to evaluate their performance under different operating conditions.

Power system simulation software has become increasingly sophisticated in recent years, allowing engineers to model and analyze large and complex power systems with a high degree of accuracy. It is now possible to simulate the behaviour of the power system in real-time, allowing operators to monitor and control the system in a safe and controlled environment.

In conclusion, power system simulation is an essential tool in power system engineering that allows engineers to model and analyze the behaviour of complex electrical networks and systems. It is used to optimize performance, evaluate new designs, and ensure the reliability and stability of the power system. As simulation software continues to improve, it is likely that it will become an even more important tool in the field of power system engineering.

1.6 Common Softwares for power system simulation:

Some popular software for power system simulation are:

- MATLAB
- CYME (Cybernetic Modelling Environment)
- MiPower
- E-Tap (Electrical Transient Analyzer Program)

Some other softwares are-

- PSS/E (Power System Simulator for Engineering)
- DIgSILENT PowerFactory
- PSAT (Power System Analysis Toolbox)
- ATP (Alternative Transient Program)
- OpenDSS (Open Distribution System Simulator)
- GridLAB-D (Grid Laboratory for Advanced Simulation of Distribution Systems)
- SKM Power*Tools
- PowerWorld Simulator

1.7 CYME

CYME is a software package for power system analysis and simulation. It is widely used in power system engineering and planning for its advanced features and capabilities. Some of the key features of CYME include:

1. **Comprehensive Modelling capabilities:** CYME allows users to model and simulate all components of the power system, including generators, transformers, transmission lines, distribution systems, and loads.
2. **Advanced simulation capabilities:** CYME can simulate a wide range of power system phenomena, including power flow, short-circuit analysis, transient stability analysis, and harmonic analysis.
3. **User-friendly interface:** CYME has a user-friendly interface that makes it easy for users to input data, create models, and run simulations. It also provides a variety of graphical displays to help users visualize the results of their simulations.

4. **Integration with GIS:** CYME can be integrated with Geographic Information Systems (GIS) to enable users to visualize power system data on a map and to incorporate spatial data into their models.
5. **Customizable reports:** CYME generates customizable reports that provide users with detailed information about the results of their simulations.
6. **Multi-language support:** CYME supports multiple languages, making it accessible to users around the world.
7. **Extensibility:** CYME can be extended with additional modules and plugins to add new features and capabilities, allowing users to tailor the software to their specific needs.

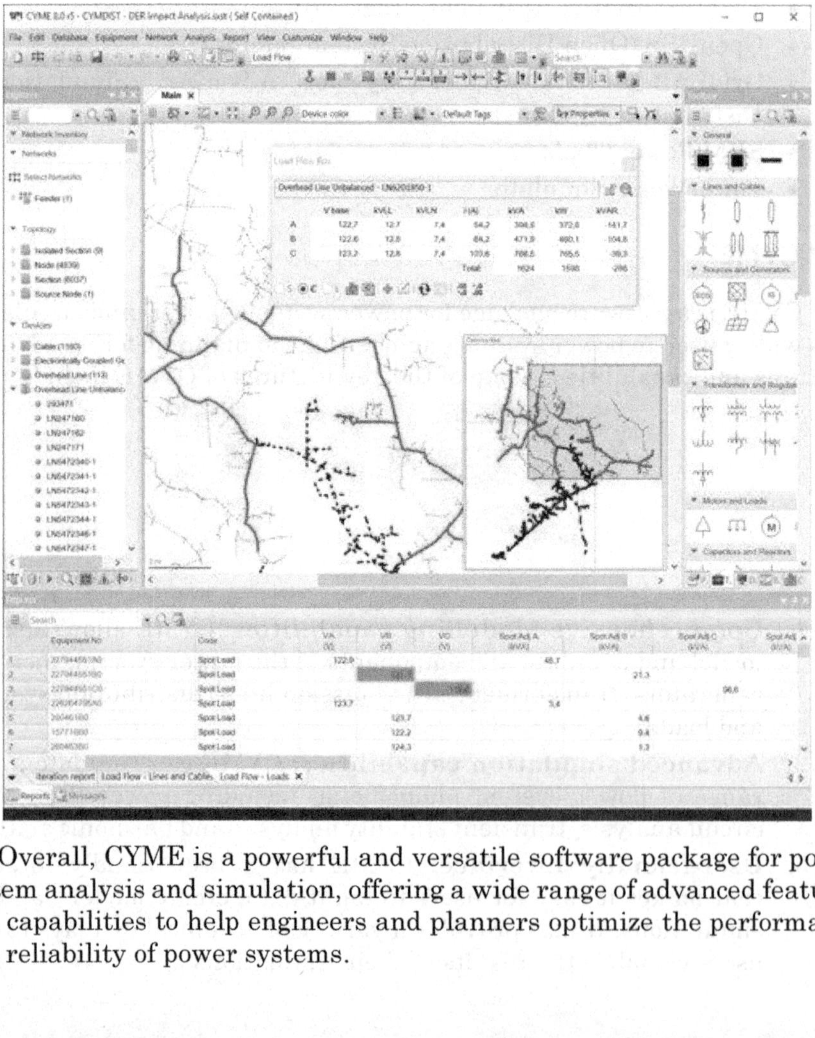

Overall, CYME is a powerful and versatile software package for power system analysis and simulation, offering a wide range of advanced features and capabilities to help engineers and planners optimize the performance and reliability of power systems.

1.8 MiPower

MiPower is a software package for power system analysis and simulation, primarily designed for the analysis and planning of distribution systems. Some of the key features of MiPower include:

1. **Comprehensive modelling capabilities:** MiPower allows users to model and simulate all components of the power system, including generators, transformers, transmission lines, distribution systems, and loads.

2. **Advanced simulation capabilities:** MiPower can simulate a wide range of power system phenomena, including load flow analysis, short-circuit analysis, voltage drop analysis, and harmonic analysis.
3. **User-friendly interface:** MiPower has a user-friendly interface that makes it easy for users to input data, create models, and run simulations. It also provides a variety of graphical displays to help users visualize the results of their simulations.

4. **Advanced optimization tools:** MiPower includes advanced optimization tools that allow users to optimize the design and operation of their power systems. These tools can be used to optimize the placement and sizing of equipment, minimize losses, and improve the overall efficiency of the system.

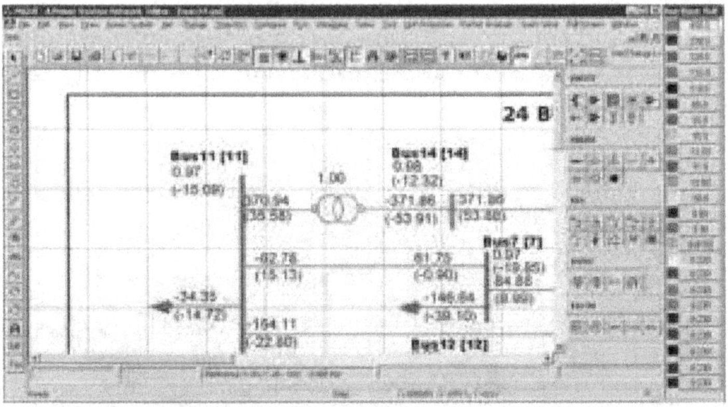

1. **Customizable reports:** MiPower generates customizable reports that provide users with detailed information about the results of their simulations. These reports can be customized to include only the information that is most relevant to the user.
2. **Integration with GIS:** MiPower can be integrated with Geographic Information Systems (GIS) to enable users to visualize power system data on a map and to incorporate spatial data into their models.
3. **Extensibility:** MiPower can be extended with additional modules and plugins to add new features and capabilities, allowing users to tailor the software to their specific needs.

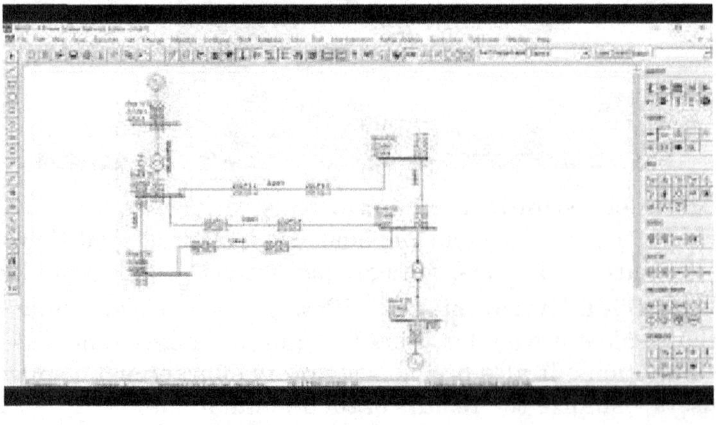

Overall, MiPower is a powerful and versatile software package for power system analysis and simulation, offering a wide range of advanced features and capabilities to help engineers and planners optimize the performance and reliability of distribution systems.

1.9 ETAP

ETAP is a comprehensive software package for power system analysis and simulation, designed for a wide range of applications in power system engineering and planning. Some of the key features of ETAP include:

1. **Comprehensive Modelling capabilities:** ETAP allows users to model and simulate all components of the power system, including generators, transformers, transmission lines, distribution systems, and loads.
2. **Advanced simulation capabilities:** ETAP can simulate a wide range of power system phenomena, including load flow analysis, short-circuit analysis, transient stability analysis, and harmonic analysis.
3. **User-friendly interface:** ETAP has a user-friendly interface that makes it easy for users to input data, create models, and run simulations. It also provides a variety of graphical displays to help users visualize the results of their simulations.
4. **Advanced optimization tools:** ETAP includes advanced optimization tools that allow users to optimize the design and operation of their power systems. These tools can be used to optimize the placement and sizing of equipment, minimize losses, and improve the overall efficiency of the system.

5. **Real-time monitoring and control:** ETAP provides real-time monitoring and control capabilities that allow operators to monitor the performance of the power system in real-time and to respond quickly to changes in operating conditions.

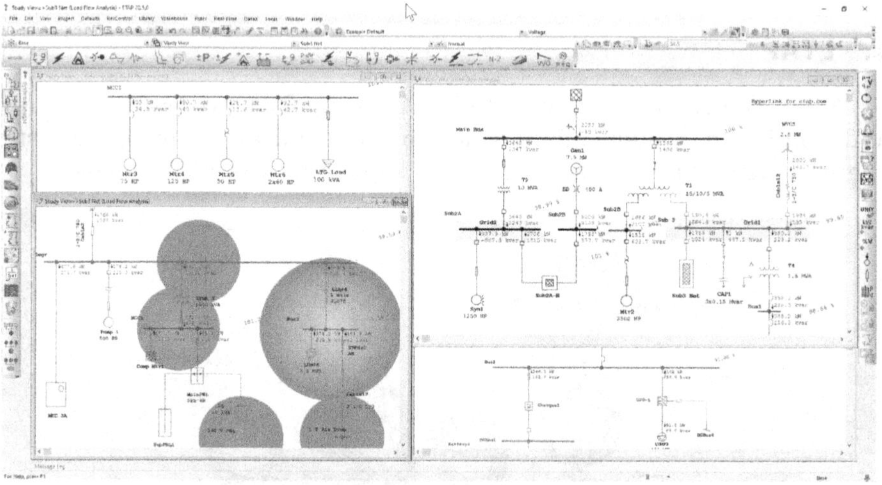

6. **Customizable reports:** ETAP generates customizable reports that provide users with detailed information about the results of their simulations. These reports can be customized to include only the information that is most relevant to the user.

Integration with other software: ETAP can be integrated with other software packages, such as GIS and SCADA systems, to enable users to incorporate additional data into their models and to monitor and control the power system more effectively.

Overall, ETAP is a powerful and versatile software package for power system analysis and simulation, offering a wide range of advanced features and capabilities to help engineers and planners optimize the performance and reliability of power systems.

1.10 MATLAB and SIMULINK

MATLAB is a numerical computing software that was developed by MathWorks in the late 1970s. The initial version of MATLAB was developed by Cleve Moler, who was a professor of computer science at the University of New Mexico. Moler developed MATLAB as a tool for teaching numerical methods to students, and the software quickly gained popularity due to its ease of use and powerful computational capabilities.

In the early 1980s, MathWorks was founded by Jack Little and Steve Bangert, with the goal of commercializing MATLAB and developing it into a full-fledged software package. Over the next few decades, MATLAB became widely adopted in various engineering and scientific disciplines, due to its ease of use, powerful computational capabilities, and extensive library of mathematical functions.

In the late 1990s, MathWorks introduced Simulink, a graphical simulation and Modelling tool that allowed users to model and simulate complex systems. Simulink quickly gained popularity among engineers and scientists, and it became an integral part of the MATLAB software package.

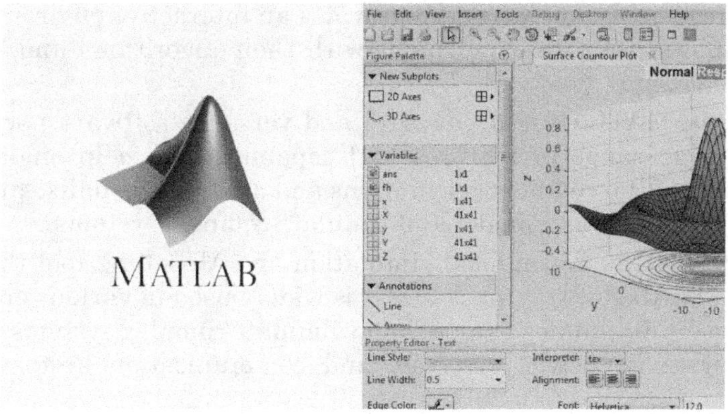

Over the years, MATLAB has continued to evolve and expand, with new features and capabilities being added with each new release. Today, MATLAB is widely used in various fields, including engineering, physics, finance, and biology, among others. Its ease of use, extensive library of mathematical functions, and powerful computational capabilities continue to make it a popular choice for engineers and scientists around the world.

MATLAB is a numerical computing software that is widely used in various engineering and scientific disciplines. Some of the key features of MATLAB include:

1. **High-level programming language:** MATLAB provides a high-level programming language that is easy to learn and use, making it a popular choice for engineers and scientists who want to quickly prototype and test their algorithms.
2. **Advanced data visualization tools:** MATLAB provides advanced data visualization tools that enable users to visualize and explore complex data sets.
3. **Comprehensive library of mathematical functions:** MATLAB includes a comprehensive library of mathematical functions, including linear algebra, optimization, and signal processing, that enable users to perform complex calculations and analysis.

4. **Extensibility:** MATLAB can be extended with additional modules and toolboxes to add new features and capabilities, allowing users to tailor the software to their specific needs.
5. **Integration with other software:** MATLAB can be integrated with other software packages, such as Simulink, to enable users to model and simulate complex systems.
6. **Platform-independent:** MATLAB is platform-independent, meaning that it can run on a wide range of operating systems, including Windows, macOS, and Linux.
7. **Interactivity:** MATLAB provides an interactive environment that allows users to experiment with their algorithms and models in real-time.

Overall, MATLAB is a powerful and versatile software package that offers a wide range of features and capabilities to help engineers and scientists perform complex calculations and analysis, visualize and explore complex data sets, and model and simulate complex systems.

SIMULINK is a graphical simulation and Modelling tool that is part of the MATLAB software package. It is widely used in various engineering and scientific disciplines to model and simulate complex systems, including control systems, power systems, and communication systems, among others.

SIMULINK provides a graphical user interface that allows users to build models using blocks that represent different components of the system being modelled. These blocks can be connected together to create a complete model of the system, which can then be simulated using MATLAB's computational engine.

SIMULINK includes a wide range of blocks for Modelling different types of systems, including linear and nonlinear systems, time-varying systems, and systems with multiple inputs and outputs. It also includes blocks for implementing different types of controllers, such as PID controllers and state-space controllers.

One of the key features of SIMULINK is its ability to simulate the behaviour of the system over time. Users can specify the inputs to the system and simulate its behaviour under different operating conditions. The results of the simulation can then be visualized using a variety of graphical displays, including time-domain plots, frequency-domain plots, and phase-plane plots, among others.

SIMULINK also includes a range of analysis tools for analyzing the behaviour of the system, including linearization tools for linearizing nonlinear systems, and optimization tools for optimizing the performance of the system.

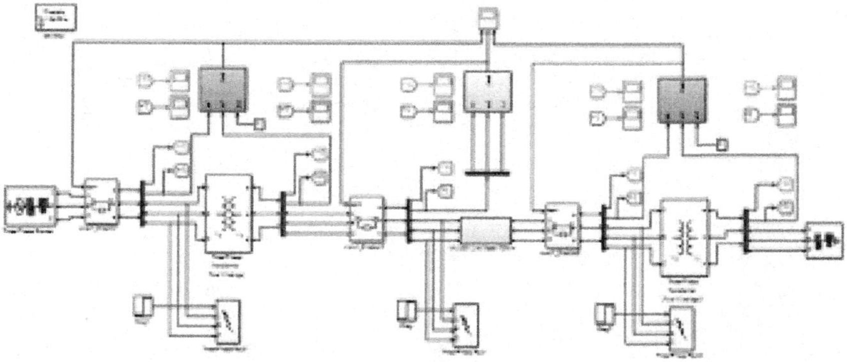

Overall, SIMULINK is a powerful and versatile tool for Modelling and simulating complex systems, providing a graphical user interface and a wide range of blocks and analysis tools to help users build and analyze models of their systems.

1.11 General Steps for Using MATLAB SIMULINK

- Open MATLAB and select SIMULINK from the MATLAB toolbar.
- Create a new SIMULINK model by selecting File > New > Model from the SIMULINK menu.
- Drag and drop blocks from the SIMULINK Library Browser onto the model canvas to create a block diagram. The Library Browser contains a variety of blocks for different applications, such as signal processing, control systems, and power systems.
- Connect the blocks with signal lines to create a system diagram.
- Configure the block parameters by double-clicking on each block to open its dialog box.
- Set the simulation parameters by selecting Simulation > Configuration Parameters from the SIMULINK menu. In the Configuration Parameters dialog box, you can set the simulation time, solver type, and other simulation options.
- Run the simulation by clicking the "Play" button on the SIMULINK toolbar or selecting Simulation > Run from the SIMULINK menu.
- Analyze the simulation results using the Simulation Data Inspector or by exporting the data to MATLAB for further analysis.
- Save the model and the simulation results for future use.

These are just general steps, and the specifics may vary depending on your particular application and model.

1.12 General Steps for using MATLAB SIMSCAPE

- Open MATLAB and select SIMULINK from the MATLAB toolbar.
- Create a new SIMULINK model by selecting File > New > Model from the SIMULINK menu.
- From the SIMULINK Library Browser, drag and drop a “Physical System” block onto the model canvas.
- Select “Add Simscape Library” from the context menu of the “Physical System” block. This will add the Simscape library to your model.
- From the Simscape library, drag and drop blocks that represent physical components such as resistors, capacitors, and mechanical systems into the model.
- Connect the blocks with physical connections such as mechanical translational and rotational joints or electrical conserving ports.
- Configure the block parameters by double-clicking on each block to open its dialog box.
- Set the simulation parameters by selecting Simulation > Configuration Parameters from the SIMULINK menu. In the Configuration Parameters dialog box, you can set the simulation time, solver type, and other simulation options.
- Run the simulation by clicking the “Play” button on the SIMULINK toolbar or selecting Simulation > Run from the SIMULINK menu.
- Analyze the simulation results using the Simscape Results Explorer or by exporting the data to MATLAB for further analysis.
- Save the model and the simulation results for future use.

These are just general steps, and the specifics may vary depending on your particular application and model.

1.13 MATLAB SimPowerSystems

MATLAB SimPowerSystems (formerly known as SimPowerSystems and SimElectronics) is a tool for simulating power electronic systems and power systems. Here are some of the tools available in MATLAB SimPowerSystems:

- **Power System Blocks:** MATLAB SimPowerSystems provides a library of power system blocks such as transformers, transmission lines, loads, generators, and protective relays that can be used to model and simulate electrical power systems.
- **Power Electronics Blocks:** MATLAB SimPowerSystems also provides a library of power electronics blocks such as converters, inverters, and rectifiers that can be used to model and simulate power electronic systems.
- **Circuit Simulation:** MATLAB SimPowerSystems can simulate electrical circuits and power electronics circuits. It provides circuit simulation tools for analyzing circuit performance and optimizing circuit designs.
- **Power System Analysis:** MATLAB SimPowerSystems can perform power system analysis, such as load flow analysis, short-circuit analysis, and transient stability analysis. It also provides tools for optimizing power system performance.
- **Control System Design:** MATLAB SimPowerSystems can design and simulate control systems for power electronics and power systems. It provides control system design tools for Modelling and simulating control systems and optimizing control system performance.
- **Visualization Tools:** MATLAB SimPowerSystems provides visualization tools for analyzing simulation results. These tools can be used to view waveforms, power spectral densities, and other simulation results.

These are just a few of the tools available in MATLAB SimPowerSystems. There are many other tools and features that can be used for power electronics and power system simulation and analysis.

1.14 Challenges in Simulation

Simulation is a powerful tool for Modelling complex systems and predicting their behaviour. However, there are several challenges in simulation that need to be addressed. One of the primary challenges is ensuring that the simulation accurately reflects the real-world system being modeled. This requires a deep understanding of the system and the ability to incorporate all relevant variables and interactions.

Another challenge is the computational power required to run simulations. As systems become more complex, the amount of computational power required to accurately model them increases exponentially. This can be a limiting factor for many organizations, particularly those with limited resources.

Finally, simulations are only as good as the data that is used to inform them. Obtaining accurate and comprehensive data can be difficult, particularly in complex systems with many variables and interactions. This can lead to inaccuracies in the simulation results and limit their usefulness.

1.15 Limitations of Simulation

While simulation is a powerful tool for Modelling complex systems, there are several limitations that need to be considered. One of the primary limitations is the accuracy of the simulation. Even with careful Modelling and consideration of all relevant variables, there will always be some degree of error or uncertainty in the simulation results. This can be due to inaccuracies in the data, limitations of the simulation software, or other factors.

Another limitation is the scalability of simulation. While simulations can be used to model small-scale systems with a high degree of accuracy, it becomes increasingly difficult to accurately model larger systems. This is due to the exponential increase in the number of variables and interactions that must be accounted for as the system grows in size.

Furthermore, simulations are often based on simplified models of complex systems. These simplifications can lead to inaccuracies and limit the usefulness of the simulation results. In addition, simulations may not account for rare events or unforeseen circumstances that can occur in the real world, which can limit the accuracy of the simulation.

Finally, simulations are limited by the resources available to run them. High-fidelity simulations require significant computational power and can be time-consuming to run. This can limit the frequency and scope of simulations that can be performed, which can in turn limit their usefulness for decision-making purposes. Overall, while simulations can be a valuable tool, it is important to be aware of their limitations and use them appropriately.

1.16 Emerging Trends and Future Direction of Simulation

Simulation has become an integral part of engineering design and analysis, and there are several emerging trends and future directions that are shaping the field. Some of these trends include advances in computational power, the integration of simulation into the product development cycle, and the increasing use of virtual and augmented reality in simulation.

Advances in computational power are enabling engineers to perform more complex simulations with higher fidelity than ever before. This is particularly true in fields such as computational fluid dynamics (CFD) and finite element analysis (FEA), where complex simulations can require massive amounts of computational resources. With the advent of cloud computing and parallel processing, engineers can now perform simulations that were once impossible due to the limitations of desktop computing. This trend is expected to continue, with even more powerful computing resources becoming available in the future.

Another emerging trend is the integration of simulation into the product development cycle. Traditionally, simulation was performed after a product design was completed, as a way to validate the design and ensure it met performance requirements. However, more and more companies are now using simulation early in the design process, as a way to explore design alternatives and optimize product performance. This approach, known as simulation-driven design, is becoming increasingly popular as companies look for ways to reduce development time and costs.

Virtual and augmented reality are also becoming more prevalent in simulation. These technologies allow engineers to interact with simulations in more intuitive ways, providing a more immersive and engaging experience. For example, virtual reality can be used to visualize a simulation in 3D space, allowing engineers to walk around and interact with the simulation as if it were a physical object. Augmented reality, on the other hand, can be used to overlay simulation data onto the real world, providing engineers with real-time feedback on how a design will perform in the real world.

In addition to these emerging trends, there are several future directions that are likely to shape the field of simulation in engineering. One of these is the increasing use of machine learning and artificial intelligence in simulation. Machine learning can be used to automate many aspects of the simulation process, such as meshing and model setup, freeing up engineers to focus on higher-level tasks such as interpreting simulation results and making design decisions. Similarly, artificial intelligence can be used to optimize simulation parameters and guide the simulation process, leading to more efficient and effective simulations.

Another future direction is the increasing use of simulation for predictive maintenance. By monitoring equipment and systems in real-time and performing simulations based on sensor data, engineers can predict when maintenance will be required and take preventative action before a failure occurs. This approach can help reduce downtime and maintenance costs, while also improving equipment reliability and performance.

Finally, simulation is likely to play an increasingly important role in the development of autonomous systems. Autonomous vehicles, drones, and robots all require sophisticated control algorithms to operate safely and effectively. Simulation can be used to test and refine these algorithms in a virtual environment, before deploying them in the real world. This approach can help accelerate the development of autonomous systems and ensure their safety and reliability.

In conclusion, simulation is an ever-evolving field that is constantly being shaped by emerging trends and future directions. Advances in computational power, the integration of simulation into the product development cycle, and the increasing use of virtual and augmented reality are just a few of the trends that are shaping the field today. Looking to the future, the increasing use of machine learning and artificial intelligence, the development of predictive maintenance systems, and the role of simulation in the development of autonomous systems are likely to be key directions for the field in the years ahead.

2

EXAMPLES AND CASE STUDIES IN POWER SYSTEM SIMULATION

SECTION A : EFFECT OF UNIFIED POWER QUALITY CONDITIONER IN SMART GRID

CHAPTER - 1
INTRODUCTION: ELECTRICAL POWER QUALITY AND SMART GRID

1.1 Electric Power Quality

Few years back it was considered that the main concern of electricity consumers was the uninterrupted power and reliability of supply. The power system reliability is defined as the continuity of electric supply. Even though the power generation in most advanced countries is fairly reliable, the distribution is not always so. The transmission systems multifarious the problem further as they are exposed to the impulses of Mother Nature. It is however not only reliability that the consumers want these days, power quality is also very significant to them. if connected to the same bus that a consumer is supplies a large motor-load may observe an acute dip in his supply voltage each time the motor-load is powered on. In some extreme cases, he may have to bear with power cuts. This may be quite undesirable to most of the energy customers.

There are also very profound loads such as air traffic control, hospitals (operation theatre, life support, patient database system), continuous process industries (chemical, semiconductor, refinery, food, rayon & fabrics), all financial organizations, data centers, and numerous other data processing and service providers that require clean and uninterrupted power. In many processes for example continuous processing industry or semiconductor industry, product manufacturing process can be spoiled by a voltage dip of very short interval. Such customers are very distrustful of such dips since each such interruption cost them a substantial amount of money. While small voltage drops are adequate to effect contactors on motor drives to drop out. Stoppage during a portion of a method will destroy the conditions for internal control of the product and need restarting of production. so during this modified situation during which the customers more and more demand quality power, the term power quality (PQ) attains enhanced significance.

Power Transmission lines are subjected to the natural forces, moreover, every transmission line has its loadability limit that's usually determined by either stability issues or by thermal limits. although the power quality drawback could be a distribution aspect problem, transmission lines usually have an effect on the quality of power supplied. it's but to be noted that whereas most issues related to transmission systems arise because of the forces of nature or because of the interconnection of power systems, individual customers are accountable for additional substantial fraction of the issues of power distribution systems

Effect of Power Quality Problems on End Users

Reason of power quality issues are typically complicated and tough to observe. Technically speaking, the best ac line offer by the utility system ought to be a pure wave of fundamental (50/60 Hz). additionally, the peak of the voltage ought to be of rated worth. sadly, the actual ac line supply that we have a tendency to receive everyday departs from the perfect specifications. Table 1.1 lists numerous power quality issues, their characterization ways and potential causes.

There are some ways within which the shortage of quality power affects customers. Impulsive transients don't travel terribly far away from their purpose of entry. but associate impulsive transient will make to an oscillating transient. The oscillating transient will result in transient overvoltage and resultant harm to the power line insulators. Impulsive transients are sometimes suppressed by surge arresters.

Short length voltage variations have varied effects on customers. Voltage sags (also is aware of as dips) will cause loss of production in automatic processes since a voltage sag will trip a motor or cause its controller to malfunction. For semiconductor producing industries such a loss is substantial. A voltage sag also can force a computing system or data processing system to crash. to stop such a crash, an uninterruptible power supply (UPS) is usually used, which, in turn, could generate harmonics.

The protecting circuit of associate adjustable speed drive (ASD) will trip the system throughout a voltage swell. conjointly voltage swells will place stress on computers and plenty of home appliances, thereby shortening their lives. a temporary interruption lasting a number of seconds will cause a loss of production, erasing of pc knowledge etc. the price of such a disruption throughout peak hours is many thousands of rupees.

Table 1.1: Power quality problems and their causes

Broad Categories	Specific Categories	Methods of Characterization	Typical Causes
Transients	Impulsive	Peak magnitude, rise time and duration	Lightning strike, transformer energization, capacitor switching
	Oscillatory	Peak magnitude, frequency components	Line or capacitor or load switching.
Short duration voltage variation	Sag	Magnitude, duration	Ferroresonant transformers, single line-to-ground faults
	Swell	Magnitude, duration	Ferroresonant transformers, single line-to-ground faults
	Interruption	Duration	Temporary (self-clearing) faults
Long duration voltage variation	Undervoltage	Magnitude, duration	Switching on loads, capacitor deenergization
	Overvoltage	Magnitude, duration	Switching off loads, capacitor energization
	Sustained interruptions	Duration	Faults
Voltage imbalance		Symmetrical components	Single-phase loads, single-phasing condition
Waveform distortion	Harmonics	THD, Harmonic spectrum	Adjustable speed drives and other nonlinear loads
	Notching	THD, Harmonic spectrum	Power electronic converters
	DC offset	Volts, Amps	Geo-magnetic disturbance, half-wave rectification
Voltage flicker		Frequency of occurrence, modulating frequency	Arc furnace, arc lamps

The impact of long period voltage variations is larger than those of short period variations. A sustained overvoltage lasting for few hours will cause harm to electrical appliances while not their owner knowing it, till it's too late. The low voltage has identical result as that of a voltage sag. within the case of a sag the termination of method is immediate. The system returns to normal mode after restoring the normal voltage. .

However, within the case of a sustained under voltage, the process cannot even be started or resumed. A sustained interruption is typically caused by faults. Since the loss to customers because of any sustained interruption are often within the order of uncountable rupees, it's necessary for the utility to possess an honest preventive maintenance schedule and to comply agreements or laws/rules to encourage high power system reliability.

Voltage imbalance will create over-heating in motors and may even results in an oversized motor to trip. Harmonics, dc offset and notching creates distortions in sinusoidal wave shape. Normally the system harmonics are integer multiples of fundamental wave or fundamental frequency fractions (called -sub harmonics). Also at some frequencies that don't seem to be integer multiples of the fundamental component (called-interharmonics). Undesired harmonic currents flowing from the power distribution lines will causes unnecessary losses.

Harmonics may also create maloperation of control systems, heating and losses in transformers, ripple management, or electro-magnetic interference [-EMI] and interference with the nearby telecommunication systems. Ripple management refers to the utilization of a 300Hz to 2500Hz signal additional to distribution lines to regulate change of loads like predicament heaters or street lighting. Interharmonic voltages will upset the operation of tubelights, computers and TV receivers. they will conjointly turn out acoustic noise in power equipment. DC offsets will cause saturation within the power transformer magnetic circuits. A notch is defined as a cyclic transient that overlaps on the supply voltage. It will harm capacitive parts connected in shunt because of high rate of voltage rise at the notches. The main reasons of voltage flickers are: -starting of big motors, arc discharge lamps, arc welding machines, arc furnaces etc. Voltage flickers creates frequent variations in supply voltage which will create the intensity of light from lightning lamps to vary. This variation is perceived as distressful by the user human observers, significantly within the vary of three to fifteen times per second. The voltage flicker will have very adverse effects on human health because the high frequency flickering of tube lights, lighting bulbs, or TV screen will cause strain upon the human eyes leading to headaches or migraines. The voltage flicker may scale back the manufacturing of electrical equipment, lamps etc. So it is concluded that the dearth of ordinary quality power can cause loss of production, harm of apparatus or appliances or will even be prejudices to human health. Therefore, it's so imperative that a high standard of power quality must be adopted. This has been proven that the power conditioning devices based on modern power electronic can be successfully employed to enhance the power quality supplied to customers.

1.1.2 Power Quality Standards

Many standards on power-quality have been expertly framed by The International Electrotechnical Commission [IEC] and the Institute of Electrical and Electronic Engineers [IEEE] . A review of various standards is given in [1]. Table 1.2 shows some of these standards that are given in [1]. We shall discuss various index for the measurements of the different power quality components later.

Table 1.2: Some power quality standards of IEC and IEEE

Phenomena	Standards
Classification of power quality	IEC 61000-2-5: 1995 [2], IEC 61000-2-1: 1990 [3] IEEE 1159: 1995 [4]
Transients	IEC 61000-2-1: 1990 [3], IEEE c62.41: (1991) [5] IEEE 1159: 1995 [4J, IEC 816: 1984 [6J
Voltage sag/swell and interruptions	IEC 61009-2-1: 1990 [3J, IEEE 1159: 1995 [4]
Harmonics	IEC 61000-2-1: 1990 [3], IEEE 519: 1992 [7] IEC 61000-4-7: 1991 [8]
Voltage flicker	IEC 61000-4-15: 1997 [9]

1.1.3 Power Quality Monitoring

Variations in Power quality are classified as either (1) disturbances or (2) steady-state variations. Disturbances pertain to abnormalities within the system voltages or currents because of some abnormal operations or fault in power system . The Steady-state variations confer with changes in RMS value from the harmonics or fundamental quantities. Normally these are monitored by special instruments such as harmonic analyzers,, voltage recorders, disturbance analyzers etc. But with the advancement within the instrumentation and sensor technology, accurate, quicker and additional precision instruments will currently be available for power quality observation and analysis.

The input data for any power quality monitor is taken by transducers. These incorporate measuring instruments such as potential transformers, current-transformers, Hall effect current and voltage transducers etc. The Disturbance monitors and Disturbance analyzers are instruments that are commonly used for the measurements of power quality parameters . There are 2 classes of those devices - (i) standard analyzers and (ii) graphics-based analyzers. the standard analyzers give data like magnitude and period of sag/swells, under/overvoltages etc. To specify the real time information Graphic based analyzers with memory are used . The advantage of such device is that the saved information are often analyzed and later to work out the source and reason for the power quality issues. additionally, these analyzers also can diagrammatically present the real time information.

Harmonic information is analyzed by means of instruments- harmonic or spectrum analyzers, which might diagrammatically show harmonic information. These are typically digital signal processor [DSP] based mostly data analyzers which will sample real time information and then perform fast Fourier transformation (FFT) to find the amplitudes and phase angles of the harmonic elements. These digital analyzers at the same time measure the currents and voltage and thus specified harmonic power are often calculated. These instruments can additionally take sampling of the signals at an awfully high rate so that specified harmonics up to concerning fiftieth order will be determined. Conjointly this is considered that the amplitudes of the upper-order harmonics are usually abundantly lower than the amplitudes of the lower-order harmonics. thus for the purpose of signal conversion and detection of the higher order harmonics, these analyzers have inbuilt high resolution analog/digital converters.

Currently, dedicated power quality measurement instruments are made which will mix each the functions of disturbance and harmonic measurements. These are graphical instruments which will additionally transmit information over communication lines. Flicker observance is completed through IEC flicker mete which records the instantaneous flickering voltage. This is defined as the instantaneous flicker level [IFL]. The recorded value of IFL is then stored and statistical computations on these data are done to work out short-term (10 min) flicker severity index and long-run flicker severity index.

1.2 The Smart Grid

1.2.1 Introduction

Today's electrical grid is designed to work as a vertical structure consisting of generation, transmission, and distribution and supported with controls and devices to maintain reliability, stability, and efficiency. However, system operators are currently facing new challenges together with the penetration of ICT within the heritage system, fast technological modification, and differing types of market players and end users. succeeding iteration, the sensible grid, are going to be equipped with communication technology systems and real time monitoring techniques to boost resiliency and supply/demand forecasting in addition on defend against internal and external threats. the development framework and structure of the smart grid is predicated upon unbundling and restructuring of the power sector and by optimizing its assets.

1.2.2 Today's Grid Versus the Smart Grid

As mentioned above, many factors contribute to the inability of today's grid to efficiently meet the demand for reliable power supply. Table 1.2 below compares the characteristics of today's grid with the preferred characteristics of the smart grid.

Table 1.3: Comparison of Today's Grid vs. Smart Grid

Preferred Characteristics	Today's Grid	Smart Grid
Active Consumer Participation	Consumers are uninformed and do not participate	Informed, involved consumers— demand response and distributed energy resources
Accommodation of all generation and storage options	Dominated by central generation—many obstacles exist for distributed energy resources interconnection	Many distributed energy resources with plug-and-play convenience focus on renewables
New products, services, and markets	Limited, poorly integrated wholesale markets; limited opportunities for consumers	Mature, well-integrated wholesale markets; growth of new electricity markets for consumers
Provision of power quality for the digital economy	Focus on outages— slow response to power quality issues	Power quality a priority with a variety of quality/price options—rapid resolution of issues
Optimization of assets and operates ef?ciently	Little integration of operational data with asset management— business process silos	Greatly expanded data acquisition of grid parameters; focus on prevention, minimizing impact to consumers

Anticipating responses to system disturbances (self-healing)	Responds to prevent further damage; focus on protecting assets following a fault	Automatically detects and responds to problems; focus on prevention, minimizing impact to consumers
Resiliency against cyber attack and natural disasters	Vulnerable to malicious acts of terror and natural disasters; slow response	Resilient to cyber attack and natural disasters; rapid restoration capabilities

- Accommodating renewables
- Power transfers across regions and Handling uncertainties in schedules
- Resolving and Managing unpredictable events and uncertainties in operations & planning more aggressively.
- Optimizing the power transfer capability of the transmission and distribution networks & meeting the demand for increased quality and reliable supply

Main Components of Smart Grid

- Real Time Monitoring and Control (SCADA Systems)
- Energy Management Systems (EMS) for Transmission Networks
- Distribution Management Systems (DMS) for Distribution Networks
- Communication Networks
- Analytics and Decision Support
- Smart Meters and AMI Infrastructure
- Building Energy Management Systems
- Energy efficient and Smart Appliances

1.2.3 Energy Independence and Security: Rationale for the Smart Grid

The India government established a sensible Smart Grid Task Force [SGTF], whose mission is "to insure coordination, awareness and integration of the various activities of the MoP and all where within the linked with smart-grid technologies and measures" [1]. The S.G. task force's activities embody analysis and development; development of wide accepted smart grid standards (through BIS) and open protocols; the connection of smart grid technologies and working process to DISCOMs regulators; the connection of smart grid technologies and regulations of grid security, system reliability, and infrastructure development, and also the relationship of smart grid technologies and practices to different sides of electricity supply, demand, transmission, distribution, economy and policy. In response to the activity of SGTF, the research and development organisations is actively engaged in:

- Smart grid research and development program
- Policy and motivation to encourage smart grid technology support for generation, transmission and distribution
- Development of infrastructure to enable smart grid deployment
- Development of widely accepted smart grid standards and protection
- Certainty of system reliability and security

As mentioned above, there are 5 important aspects of smart grid deployment and development.

1.2.4 Computational Intelligence of Smart Grid

The "Computational intelligence" is defines as a term accustomed to express the advanced analytical tools required to optimize the major power network. The toolbox will include heuristic, evolution programming, decision support tools, and adaptive optimization techniques.

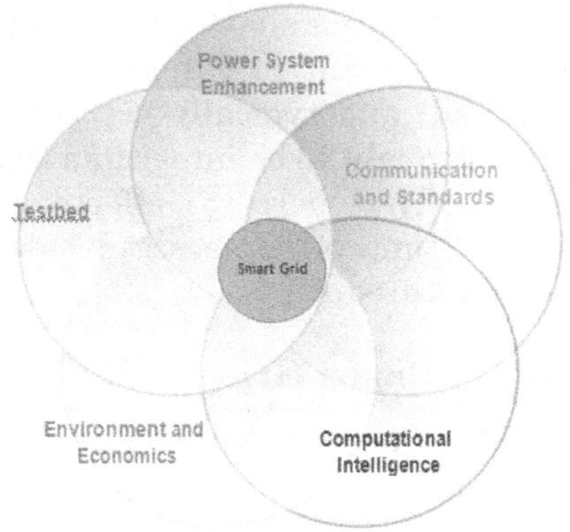

Figure 1.1: 5 key Features of smart grid development.

1.2.5 Power System Enhancement

Policy makers believe that bulk use of renewable energy resources in the India will help to reduce the impacts of carbon emissions from thermal and fossil energy, balance demand uncertainty, and to some extent, increase economy and reliability of energy delivery.[4,5]

1.2.6 Communication and Standards

Since planning horizons can be short as 45 minutes ahead, the smart grid's advanced automations will generate vast amounts of operational data in a fast decision making environment. New algorithms will help it become adaptive and capable of predicting with foresight. In turn, new rules will be needed for managing, operating, and marketing networks.

1.2.7 Environment and Economics

Based on these desired options, associate assessment of the variations within the characteristics of the existing grid and also the projected smart grid is required to focus on characteristics of the grid and also the challenges. once totally developed the sensible grid system can enable consumers involvement, enhance generation and transmission with tools to permit minimisation of power system vulnerability, reliability, adequacy,resiliency, and power quality. The training tools and capability development to manage and operate the grids and thus crate new job opportunities is an element of the required goals of the smart grid evolution which is able to be tested using test-bed. to attain the fast preparation of the grids test bed and research centers have to be compelled to work across disciplines to create the first generation of smart grid.

By concentrating on security controls instead of individual vulnerabilities and threats, utility and smart-grid technology vendors will rectify the basis causes that result in vulnerabilities. However, security controls are a lot of difficult and typically not possible to feature to an existing system, and ideally ought to be integrated from the start to reduce implementation problems. The operative effectiveness of the enforced security controls base are going to be assessed habitually to protect the smart grid against evolving threats.

1.2.8 General View of the Smart Grid Market Drivers

In order to improve reliability and efficiency of smart grid, many market drivers and new opportunities recommend that the smart grid must -:

1. Satisfies the requirement for enhanced integration of digital systems for raised efficiency of the power system. within the restructured surroundings, the deregulated electrical utility business permits a renovation of the market to be based on system constraints and therefore the seasonal and daily fluctuations in demand. Competitive markets increase the transmission of power between regions, that any strains today's aging grid and needs updated, real-time controls.

2. Handle grid congestion, increase customer participation, and cut back uncertainty for investment. this needs the improvement of the grid's capability to handle demand dependably.
3. Seamlessly integrate renewable energy systems (RES) and distributed generation. The forceful increase within the integration of cost-competitive distributed generation technologies affects the power system

The Contribution of stakeholders to the development of the smart grid exist in addition to policy makers and independent system operators, Their specific contributions and conceptual understanding of the aspects to be undertaken are discussed below:

1.2.9 Stakeholder Function and Roles

As within the old system, essential attention should be paid to the identification of the stakeholders and the way they perform within the grid's development. Stakeholders vary from utility and energy producers to industrialists, shoppers, policy makers, technology suppliers, and researchers. a very important element in the conclusion of the smart grid is that the complete buy in or involvement of all stakeholders.

Policy makers are the federal and state regulators accountable for guaranteeing the cohesiveness of policies for modernization efforts and mediating the requirements of all parties. the first benefits of smart grid development to those stakeholders issues the mitigation of energy costs, reduced dependence on foreign oil, enhanced efficiency, and reliability of power supply. Figure 1.2 shows the classes of stakeholders.

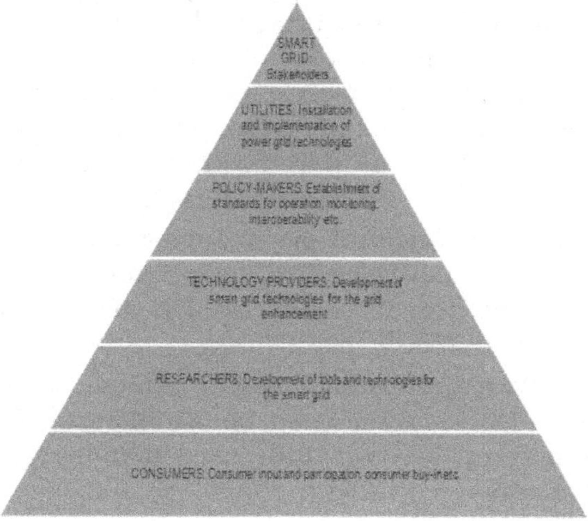

Figure 1.2: Five key aspects of smart grid development

Other participants within the development of the smart grid embody government agencies, manufacturers, and research institutes.

In the government report, named- "The smart Grid: An Introduction," - the MoP discusses the opportunities, character, challenges and necessity for smart grid enactment. It defines the smart grid as technology that "makes this transformation of the electrical industry attainable by utilizing the philosophies, ideas and technologies that enabled the web to the utility and therefore the electrical grid and permits the grid modernization" [1]. The characteristics of the good grid are bidirectional data communication, plug-and-play capabilities, advanced metering infrastructure [AMI] for integrating customers, facilities for enhanced customer involvement, interoperability supported standards, and low price communication and hardware.

Additional options identified embody integration and advancement of grid visualisation technology to produce group action real time device data, wide-area grid awareness, weather data, and grid modelling with geographical information system [1].

However, the MoPs definitions don't offer measures for addressing uncertainty, productivity, and foresight. Another federal organisation, the "Federal Energy Regulatory Commission (FERC)" of USA, by law has mandated the development of:-

1. **Cyber Security:** need agency define standard and protocol in line with the overarching cyber security and reliability necessities of the Energy Independence and Security Act (EISA) and therefore the FERC reliability Standards.
2. **Coordination of the majority power systems with new and rising technologies:** establish standards development that facilitate to accommodate the introduction and growth of renewable resources, demand response, and electricity storage to address many bulk power system challenges. additionally determine standards development that facilitate to accommodate another rising technology, electrical transportation.
3. **Wide Area Situational Awareness:** make sure that system operators must have the proper instrumentation that provides them an entire view of their systems so that they will monitor and operate their systems.
4. **Intersystem Communications:** determine standards for common info models for communication among all parts of the majority grid regional market operators, utilities, demand response aggregators, and customers

Some key characteristics of the Intelligent grid are:

- Grid equipment and assets contain or are monitored by intelligent IP-enabled devices (digital processors).
- Digital communication networks allow the intelligent devices to communicate firmly with the utility enterprise and possibly with one another.
- Data from the intelligent devices and plenty of alternative sources are consolidated to support the transformation of information into helpful info through advanced analytics.
- Business intelligence and optimization tools give advanced decision support at each the automated and human supervisory level.
- The information base and architecture contains five major components: information sources, information transport, information integration, analytics, and optimisation. additionally, there are means that for information distribution which incorporates publish-and-subscribe middleware, portals, and Web- based mostly services [8].

1.2.10 Working Definition of the Smart Grid Based on Performance Measures

An operating definition must embody the subsequent attributes:

- Assess grid health in real time
- Predict behaviour, anticipate
- Adapt to new environments like distributed resources and renewable energy resources
- Handle random demand and respond to smart appliances
- Give self-correction, reconfiguration, and restoration
- Handle randomness of loads and market participants in real time produce additional complicated interactive behaviour with intelligent devices, communication protocols, and normal and smart algorithms to boost smart communication and transportation systems.

In this surroundings, smart management methods can handle congestion, instability, or reliability issues. The smart grid is going to be cyber-secure, resilient, and able to manage shock to confirm durability and reliability. extra options embrace facilities for the integration of renewable and distribution resources, and getting data to and from renewable resources and plug-in hybrid vehicles. New interface technologies can create information flow patterns and data obtainable to investors and entrepreneurs inquisitive about making product and services. Thus, the working definition becomes:

"The smart grid is an advanced digital bidirectional power flow power system capable of self-healing, and adaptive, resilient, and sustainable, with foresight for prediction under different uncertainties. It is designed for interoperability with present and future standards of elements, devices, and systems that are cyber-secured against malicious attack".

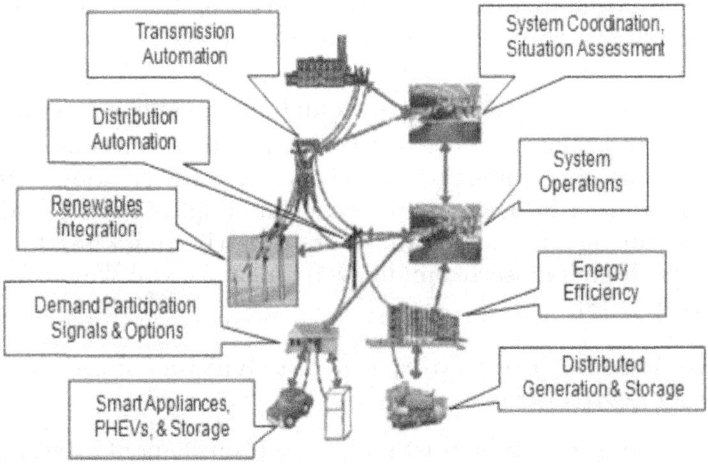

Figure 1.3: DOE representative architecture of the smart grid design

It is enabled to perform with strong and reasonable real-time measurements and advanced communication technology for data/ information transmission. It permits smart appliances and facilitates the readying of advanced storage technologies together with plug-in electrical and hybrid vehicles and control choices, and supports DSM and demand response schemes.

1.2.11 Representative Architecture

Several kinds of architecture are projected by the assorted bodies concerned in smart grid development. we tend to present two: one from the DOE and one illustrated by Figure one.5, that shows however the DOE's planned smart grid divides into 9 areas: transmission automation, system coordination situation assessment, system operations, distribution automation, renewable integration, energy efficiency, distributed generation, demand contribution choices, smart home appliances, PHEV, and storage.

Figure 1.4 shows however the second architectural framework is divided into sub- systems with layers of intelligence and technology and new tools and innovations. It involves bulk power generation, transmission, distribution, and end user level of the electrical power grid. The function of every part is explained within the next section.

1.2.12 Functions of Smart Grid Components

For the generation level of the power system, smart enhancements will extend from the technologies used to improve the stability and reliability of the generation to intelligent controls and the generation mix consisting of renewable resources.

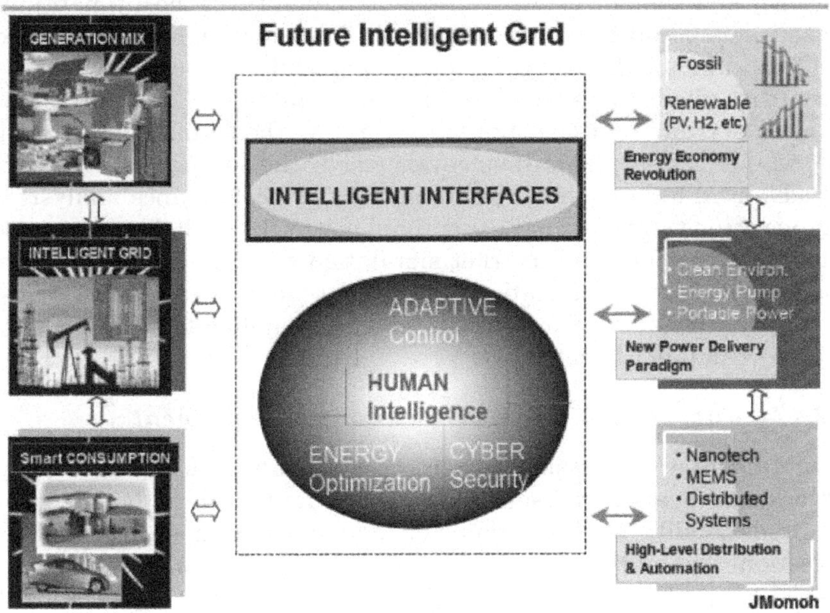

Figure 1.4: The intelligent grid (Architecture 2)

1.12.1 Smart Devices Interface Component

Smart devices for observation and control type a part of the generation components' real time data processes. These resources ought to be seamlessly integrated within the operation of each centrally distributed and district energy systems.

1.12.2 Storage Component

Due to the variability of renewable energy and also the disjoint between peak availableness and peak consumption, it's vital to find ways in which to store the generated energy for later use. choices for energy storage technologies embrace pumped-up hydro, advance batteries, Flow batteries, compressed gas, super-conducting magnetic energy storage, super-capacitors, and flywheels. Mechanisms of allied market for managing renewable energy resources, distributed generation, environmental impact and pollution are different elements necessary at the generation level.

1.12.3 Transmission Subsystem Component

In the main support structure of an integrated power grid, the power transmission system interconnects all substations and major load centres. Efficiency and reliableness at a reasonable value still be the ultimate aims of transmission planners and operators. All Transmission lines should accept variations in load change and contingency while not service disruptions. to confirm reliableness, performance, and quality of supply standards are most popular contingency parameters. Methods to attain smart grid performance at the transmission level embody the design of analytical tools and advanced technology with intelligence for performance analysis like-dynamic strong state estimation, optimal power flow, reliableness real time stability assessment, and market simulation tools. Real time monitoring supported PMU, communication technologies and state estimators sensors, are the transmission subsystem's intelligent enabling tools for developing smart transmission practicality.

1.12.4 Monitoring and Control Technology Component

Intelligent transmission systems/assets embody a smart intelligent network, self-healing and self- observance, and therefore the adaptability and predictability of generation and demand strong enough to handle congestion, reliability, and instability problems. This new strong smart grid must be capable to bear shock (reliability and durability), and be reliable to produce real-time changes in its application.

1.12.5 Intelligent Grid Distribution System Element

The distribution system is that in the final stage within the transmission of power supplies to end users. Primary feeders at this voltage level supply small industrial customers and secondary distribution feeders supply residential and industrial customers. At the distribution level, intelligent support schemes can have observance capabilities for automation exploitation good meters, communication links between customers and utility management, energy management elements, and AMI. The automation perform will be equipped with self-learning capability, together with modules for automatic billing, fault detection, load transfer, voltage improvement restoration, real-time pricing and feeder reconfiguration.

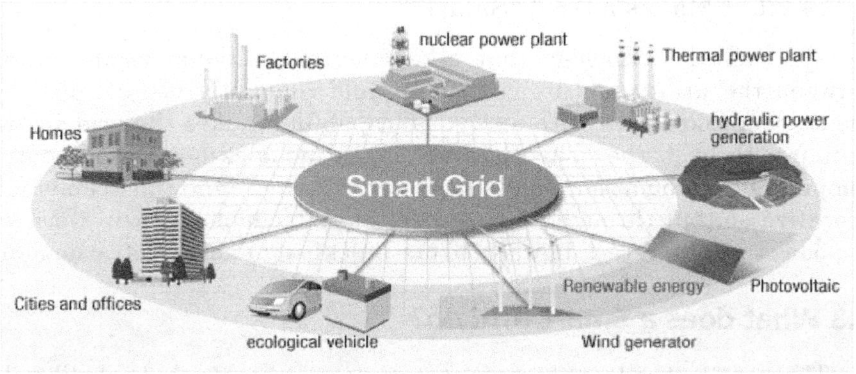

Figure 1.5: Smart grid elements

1.2.13 Demand Side Management Component

Demand side management choices and energy efficiency choices developed for effective means that of modifying the consumer demand to curtail in operation expenses from costly generators and defer capability addition.

DSM choices offer reduced emissions in fuel production, lower prices, and contribute to reliability of generation. These choices have an overall impact on the utility load curve. a regular protocol for customer delivery with bidirectional data migration technologies because the enabler is required. Plug-and-play, smart energy buildings and smart homes, demand-side meters, clean air needs, and customer interfaces for higher energy efficiency are in situ.

A smart grid is an electrical power grid which contains a multiplicity of operational and energy measures beside smart meters, smart electrical appliances, renewable energy resources, and energy economical resources etc. "The grid", refers to the electrical power grid, a network of HV/LV transmission lines, substations, transformers and a great deal of that deliver electricity from the facility plant to your home or business. It's what we tend to plug into once flip on our light switch or power up our portable computer. though the electrical grid is taken under consideration an engineering marvel, we tend to are stretching its patchwork nature to its capability. to maneuver forward, we want a replacement moderately electrical grid, one that is designed from the bottom up to handle the groundswell of digital and computerized instrumentation and technology obsessed on it—and one which can alter and manage the increasing quality and desires of electricity at intervals the twenty 1st Century.

1.2.14 What Makes a Grid "Smart?"

The digital technology that allows for bidirectional communication between the utility and its customers, and conjointly use of sensors on the transmission lines with analytical capability makes the grid smarter. similar to the web, the smart Grid technology includes monitoring, controls, computers, automation, and new technologies and instrumentality operative on, but throughout this case, these technologies will work with the power grid to retort digitally to our quickly dynamic electric demand.

1.3 What does a Smart Grid do?

The smart Grid represents associate unprecedented likelihood to maneuver the energy trade into a brand new era of reliability, accessibility, and efficiency which can contribute to our economic and environmental health. throughout the transition length, it will be vital to carry out testing, technology enhancements, consumer education, development of standards and rules, and information sharing between comes to substantiate that the benefits we've got an inclination to check from the smart Grid become a reality. the benefits associated with the smart Grid include:

- Improved system stability and security
- More economical transmission of electrical power
- Increased integration of larger scale renewable energy systems
- Better integration of customer owner power generation systems, together with renewable energy systems
- Reduced operations and management prices for utilities, and ultimately lower power prices for customers
- Quicker restoration of electricity once power disturbances
- Reduced peak demand, which is able to additionally facilitate lower electricity rates

Today, an electricity disruption like a blackout will have a domino effect—a series of failures that may have an effect on banking, communications, traffic, and security. this is often a selected threat within the winter, once householders may be left without heat. a smarter grid can add resiliency to our electric power System and build it better ready to handle emergencies like severe storms, earthquakes, massive solar flames, and terrorist attacks. due to its bidirectional interactive capability, the smart Grid can provide automatic rerouting once equipment fails or outages occur. this can minimize outages and minimize the results once they do happen.

On every occasion power outage occur, the smart grid can observe and isolate the outages, containing them before they become large scale blackouts. Latest technologies will enable to make sure that electricity recovery resumes quickly and strategically when emergency routing electricity to emergency services 1st, for instance. the smart Grid furthermore can take larger advantage of customer owned power generators to provide power once it's not available from utilities. By combining these "distributed generation" resources, a community might keep its hospital, local department, traffic lights, telephone system, and market operative throughout emergencies. furthermore, the smart Grid may be a way to address an aging energy infrastructure that has to be upgraded or replaced. It's the simplest way to handle energy efficiency, to bring enhanced awareness to consumers regarding the affiliation between electricity use and therefore the environment. And it's the simplest way to bring enhanced national security to our energy System drawing on bigger amounts of home produced electricity that's a lot of immune to natural disasters and attack.

1.3.1 Giving Consumers Control

The smart Grid isn't concerning as regards to the subject of reference to utilities and technologies; it's about supplying you with the data and tools you wish to form decisions about your energy use. If you already manage activities like personal banking from your electronic computer, imagine managing your electricity in a very similar approach. a smarter grid can improve a unique level of customer participation. for example, you'll now not need to look ahead to your monthly statement to understand what proportion electricity you employ. With a smarter grid, you'll be able to have a transparent and timely image of it. "Smart meters," and different mechanisms, can permit you to visualize what proportion electricity you employ, once you use it, and its value. Combined with time period valuation, this can permit you to avoid wasting money by using less power once electricity is costliest. While the potential advantages of the smart Grid are typically mentioned in terms of economics, national security, and renewable energy goals, the smart Grid has the potential to assist you save cash by serving to you to manage your electricity use and opt for the simplest times to get electricity. And you'll be able to save even a lot of by generating your own power.

1.4 Historical Development of the Electrical Power Grid

In Great Barrington, Massachusetts USA the 1st electrical power grid system was coming into the operation in the year 1886. At that point, the grid was a centralized radial type of electrical power transmission, electricity distribution, and based on demand management.

Developments of local electrical grids in the 20th century and were interconnected for efficiency and responsibleness. the developed countries had, terribly massive power grids in 1960, developed and extremely interconnected, with thousands of centralised interconnected generating power stations supplying electricity to major load centres through high power capability and high voltage transmission lines.

Electricity consumption must be necessary metered on every user basis so as to permit appropriate charge in keeping with the [highly variable] level of consumption of various users. due to restricted information assortment.

In some areas, offer of electricity, particularly at peak times, couldn't continue with this demand, leading to poor power quality together with blackouts, power cuts, and brownouts. more and more, electricity was trusted for business, heating, communication, lighting, and amusement, and consumers demanded ever higher levels of reliability.

By the end of the 20th century, the power demand patterns were recognized. The comparatively low utilization of those peaking generators (commonly, gas turbines were used because of their comparatively lower opportunity cost and quicker start-up times), along with the mandatory redundancy within the electricity grid, resulted in high prices to the electricity corporations, that were passed on within the variety of enhanced tariffs. within the twenty first century, some developing countries like China, India, and Brazil were seen as pioneers of smart grid deployment.

Modernization Opportunities

Power from electrical phenomenon cells (and to a lesser extent wind turbines) has conjointly, considerably, referred to as into question the imperative for giant, centralised power stations. The apace falling prices purpose to a serious modification from the centralised grid topology to at least one that's extremely distributed, with power being each generated and consumed right at the bounds of the grid. Finally, growing concern over terrorism in some countries has led to demand a lot of sturdy energy grid that's less obsessed with centralised power stations that were gave the impression to be potential attack targets.

Definition of "Smart Grid"

The first authorized description of smart Grid was delivered by the USA's Energy Independence and Security Act -2007. The bill provides an outline under section Title XIII, with 10 characteristics, which will be thought-about a definition for good Grid, as mentioned below:

"It is that the policy of the government to support the modernization of the Nation's electricity transmission and distribution system to take care of a reliable and secure electricity infrastructure which will meet future demand growth and to realize every of the subsequent, that along characterize a smart Grid:

1. Multiplied use of digital data and controls technology to boost responsibility, security, and efficiency of the electrical grid.
2. Dynamic optimization of grid operations and resources, with full cyber-security.
3. Provision to customers of timely data and management choices.
4. Development and incorporation of demand response, demand-side resources, and energy-efficiency resources.
5. Integration of `smart' appliances and consumer devices.
6. Preparationof`smart'technologies(real-time,automated,interactive technologies that optimize the physical operation of appliances and consumer devices) for metering, communications regarding grid operations and standing, and distribution automation.
7. Readying and integration of advanced electricity storage and peak-shaving technologies, as well as plug-in electrical and hybrid electrical vehicles, and thermal storage air conditioning.
8. Preparation and integration of distributed resources and generation, as well as renewable resources.
9. Identification and lowering of unreasonable or unneeded barriers to adoption of smart grid technologies, practices, and services.
10. Development of standards for communication and ability of appliances and instrumentality connected to the electrical grid, as well as the infrastructure serving the grid.

The application of ICT, information and communication technology,digital process and communications system to the power grid, creating knowledge flow and knowledge management central to the Smart grid. Large volume of capabilities result from the combined use of information and communication technology with electric power grids along with an digital layer is the essence of the smart grid and business method transformation is necessary to take advantage of the investments in smart technology.

The Smart grid technologies emerged from earlier makes an attempt at mistreatment smart metering, electronic control, and observation. within the Nineteen Eighties, AMR (automatic meter reading) was used for observation loads from large consumers, and progressed into the AMI (Advanced Metering Infrastructure) of the Nineteen Nineties, whose meters may record data however energy was consumed at completely different periods of the day. Beginning in year 2000, Telegestore Project of Italy was the primary to network massive numbers [27 million] of homes using smart meters communicated through low band width PLC communication. the Bonneville Power Administration extended its smart grid analysis with model sensors, synchronization and Monitoring of wide area networks (WAN) were revolutionized within the early Nineteen Nineties that are capable of terribly fast analysis of anomalies in electricity quality over extremely massive geographic areas. The end result of this work was the initial commissioning of a Wide-area measurement System [WAMs] in 2000. different countries are fast desegregation this technology . China initiated a national WAMS system once the past 5-year economic plan completed in 2012.

The earliest deployments of smart grids embody the Italian system Telegestore (2005), the mesh network of Austin, Lone-Star State (since 2003), and therefore the smart grid in Boulder, Colorado (2008).

1.4.1 Some Features of the Smart Grid

- Self-healing- will have sophisticated Grid monitors & controls
- Secure from physical and Cyber threats
- Support widespread use of Distributed Generation
- Efficient management of Energy by control of end user appliances

Initial power lines within the grid were engineered employing a radial model, later connectivity was secured via multiple routes, stated as a network structure. However, this created a brand new problem: if this flow or connected effects across the network exceed the limits of any explicit network component, it might fail, and therefore the current would be shunted to different network parts, that eventually might fail additionally, inflicting effect. One of the method to overcome is by load shedding [blackout] or by voltage reduction (called-brownout).

1.4.2 Smart Grid Communication Network

There are two proven technologies are used for smart grid communication

1. Zigbee and
2. 6lowpan.

Zigbee is low power radio communications similar to Wi-Fi and Bluetooth, but it has less startup time and large number of devices up to 64000 can be connected in mesh topology. This makes Zigbee as good means of communication in smart grid. There are three types of devices used in Zigbee- called as zigbee end device, zigbee router and zigbee coordinator. Zigbee has 128 bit security system.

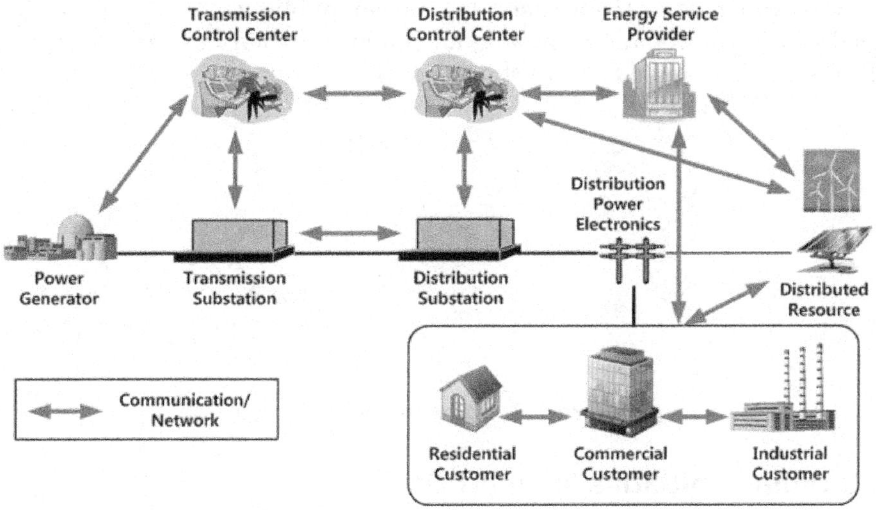

Figure 1.6: Smart grid communication network

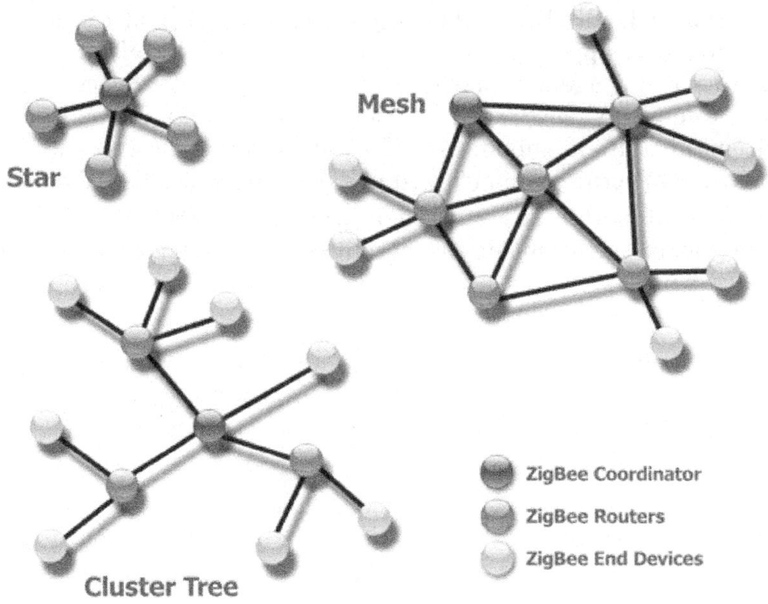

Figure 1.7: Zigbee Communication

6lopwan means IPV6 based low power public wide area network. It used IPV6 protocol and tunneling for communication from smart meter to IPV6 servers installed at smart grid control center.

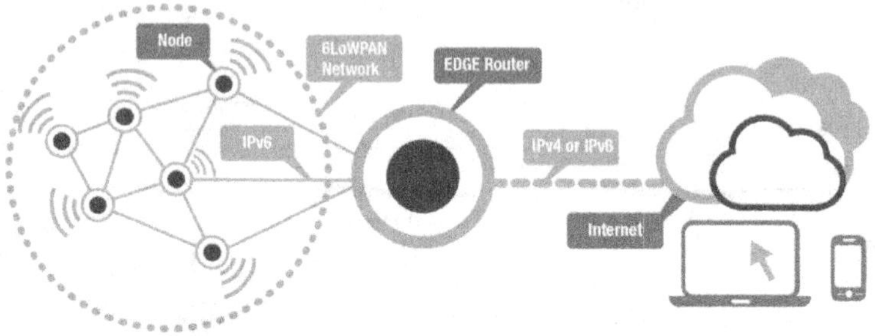

Figure 1.8: 6Lopwan Communication

1.5 India's Initiative in Smart Grid

Government of India has taken active initiative for smart grid implementation in the country under ministry of power. Smart grid forum has been formed. All DISCOMs are joined as member of the forum. The government has appointed Power Grid Corporation of India Limited (PGCIL) as nodal agency for the project. Some pilot projects have been initiated under RAPDRP part A scheme. PGCIL has introduced phasor measurement units in sub-stations for data acquisition and analysis. The organization structure of smart grid has been shown in figure below-

Ministry of power has started a website – WWW.ISGTF.IN which provides the information on the activities of smart grid task force has mentioned below—

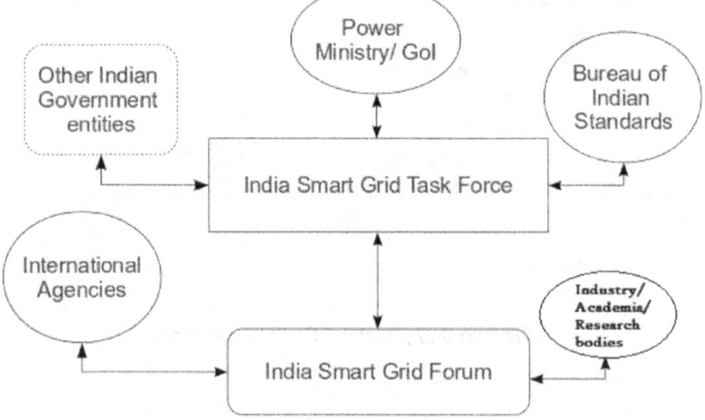

Figure 1.9: India's Smart Grid organisation

On 1st July 2011, first mini smart grid project was commissioned at Gul Pahadi, Gurgaon. The Baramati Town in Pune district of Maharashtra was the first smart grid town of India commissioned in 2014.

In the year 2014, as per the proposals received by smart grid committee 14 pilot projects were finally selected. The selection was considered on the basis of several criteria such as operational conditions, technical facilities, communication facilities, geographical factor. An amount Rs. 400 Crore was sanctioned for the smart-grid pilot project.

The national smart grid mission (NSGM) has made a document called smart grid vision for India, which proposed the digitalization of Indian power sector by the year 2027. The NSGM also made objectives for smart costumers, smart utilities, smart generation and transmission and smart policy. The mission proposed smart meter role out for the country by the year 2022.

Table 1.4: India's Smart Grid Plan

12th Plan (2012 – 2017)	13th Plan (2017 – 2022)	14th Plan (2022 – 2027)
1. Access to "Electricity for All" 2. Reduction of transmission losses (>66 kV) to below 3% 3. Reduction of AT&C losses in all Distribution Utilities to below 15% 4. Reduction in Power Cuts; Life line supply to all by 2015; grid connection of captive plants >1 MW 5. Renewable integration of 30 GW; and EV trials 6. Improvement in Power Quality and Reliability 7. ToU (Time of Use) Tariff 8. Energy Efficiency improvements 9. Standards Development for Smart Grids including EVs 10. Strengthening of EHV System 11. Efficient Power Exchanges 12. Training & Capacity Building 13. Customer Outreach & Participation 14. Research & Development 15. Sustainability Initiatives 16. SG Pilots 17. Cost-Benefit Analysis	1. Reduction of transmission losses (>66 kV) to below 2% 2. Reduction of AT&C losses to below 12% in all Utilities 3. Improvement in Power Quality 4. End of Power Cuts; Peaking power plants; Electrification of all households by 2020 5. Nationwide Smart Meter roll out 6. Renewable integration of 70 GW; 5% EV penetration 7. Standards Development for Smart Infrastructure (SEZ, Buildings, Roads/Bridges, Parking lots, Malls) and Smart Cities 8. UHV and EHV Strengthening 9. Research & Developments; Training & Capacity Building 10. Export of SG products, solutions and services to overseas 11. Customer Outreach & Participation 12. Sustainability Initiatives & Public Safety	1. Reduction of AT&C losses to below 10% in all Utilities 2. Financially viable utilities 3. Stable 24x7 power supply to all categories of consumers all across the country 4. Renewable integration of 120 GW; 10% EV penetration 5. Smart Cities and Smarter Infrastructures 6. Export of SG products, solutions and services to overseas 7. Research & Development ; Training & Capacity Building 8. Active Participation of "Prosumers" 9. Sustainability Initiatives & Public Safety

There is some barrier to smart grid applications and implementation in India. The massive investment needed for the information technology and communication technology application for converting existing grid into smart grid. The poor financial health of state owned power distribution companies is the biggest barrier. It may also cause impact of consumer tariff. There is also requirements of legislation which can make smart grid mandatory for power sector, also there is no proven technology which is viable for larger number of consumers as in India. There is need of making standard for smart grid components and technology. Adjoining legacy components based on proprietary protocol with smart grid components is also a technical challenge. There are certain communication issues in MPLS, GPRS, CDMA, ZIGBEE and 6lopwan connectivity.

Security is also a big challenge in smart grid implementation because installing smart meters at every home is the entry point for hackers.

1.6 Power Quality Problems in DG Integrated Smart Grid Network

Electric power is that the results of a production method and in line with the Council of European Energy Regulators (CEER) [9] the standard of electrical power offer ought to comprise in 3 main areas, shown in Figure two, wherever the facility quality means that the continuity of offer and voltage quality. Again, inexperienced House Gas (GHG) emission and therefore the warming are the facet effects of the traditional power production method. Therefore, developed countries also are attempting to cut back their overall GHG emission by introducing and increasing the share of renewable energy into their electrical grid system. Hence, the standard of power offer has become a crucial issue with the high penetration of dg systems either connected to the good grid or microgrid. because the star, wind, micro-hydro are the foremost leading sources of dg systems thus power quality issues associated with these DG system together with diesel (one of the best carbon dioxide emitter) are known and shown in Table two.

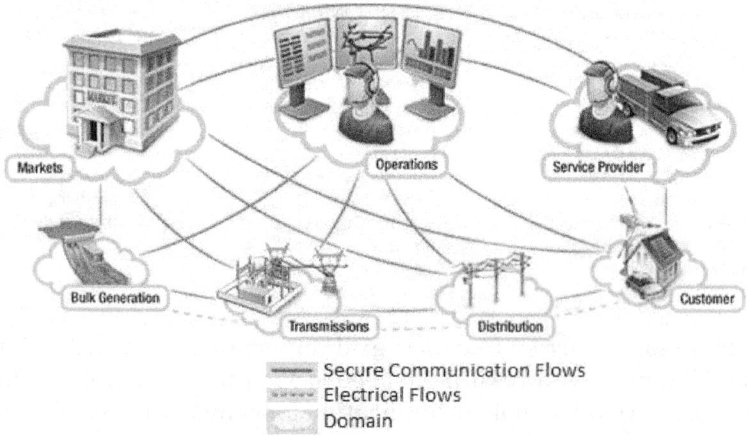

Figure 1.10: Conceptual model of Smart Grid [3]

Table 1.5: Components of Smart Grid Network

Smart Grid Components				
Bulk Generation	Transmission	Distribution	Consumer	Communication
• Smart Generation	• Sub-station Automation & Protection • Power Quality & Power monitoring System • Energy Management System • Decision Support Systems • Power Electronics	• Asset Management System and Condition Monitoring • Distribution Automation & Protection • Power Quality • Distrtibution Management System • Smart Meter	• Smart Consumption • Local Production • Power Quality • Smart House • Building Automation	• Communication • Security

Table 1.6: PQ problems related to DG systems

PQ Problems	Wind Energy	Solar Energy	Micro/ Small Hydro	Diesel
Voltage Sag/Swell	●		●	●
Over/Under Voltage	●			●
Voltage Unbalance		●		
Voltage Transient	●			
Voltage Harmonics	●	●	●	
Flicker	●	●		●
Current Harmonics	●	●	●	
Interruption	●	●		

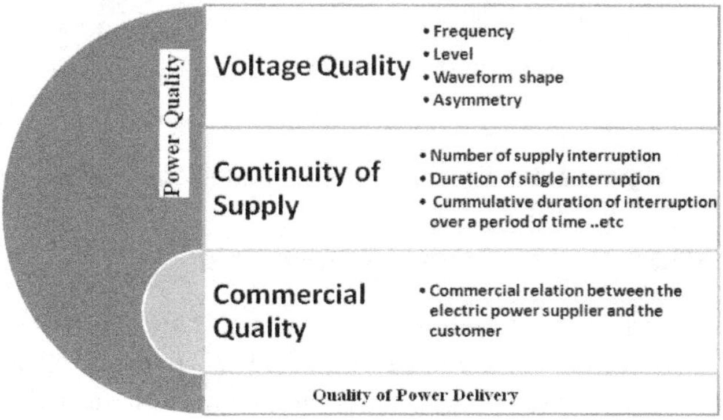

Figure 1.11: Quality of Electric Power Supply

1.7 Working Principle of UPQC

The UPQC is integration of series (APFse) and shunt (APFsh) active power filters, connected back to back on the dc side, sharing a typical DC capacitor [10], shown in Figure three. The series element of the UPQC is accountable for mitigation of the supply side disturbances- voltage sags and swells, EMF unbalancing, flicker and system harmonics. It inserts voltages to maintain the voltages at load points within the safe range alongwith balanced and distortion free. The shunt element is responsible for mitigating this quality issues caused by the consumer poor power issue, load harmonic currents, load unbalance etc. It injects currents within the ac system such the supply currents become balanced sinusoids and in phase with the supply voltages. the perform of UPQC function depends on the series and shunt APF controller. A basic purposeful diagram of a UPQC controller is shown in Figure four. the shunt APF injects the compensating reactive and harmonic current using hysteresis current controller and wherever because the series APF uses PWM voltage controller to attenuate the voltage disturbances.

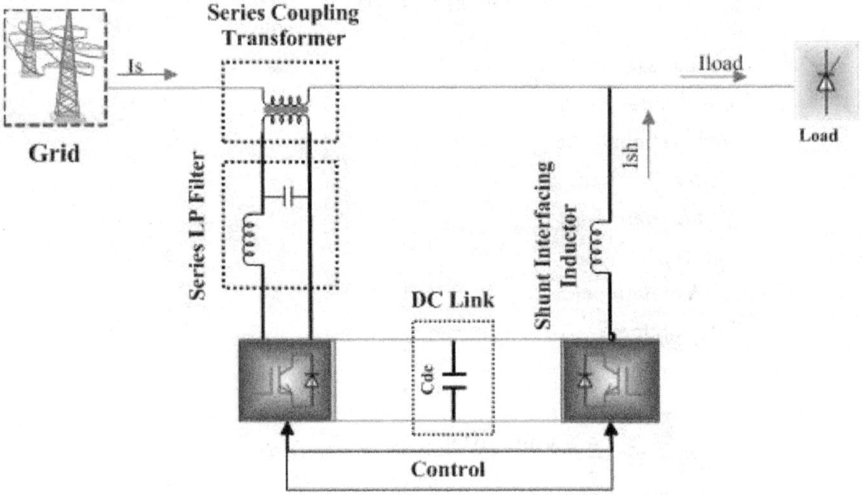

Figure 1.12: Basic System Configuration of UPQC

1.8 Integration of UPQC with Smart Grid

Latest reports [4-8, 11, 12] show that important analysis and development has been administrated on the applying of UPQC to dg integrated network. because the UPQC will catch up on the majority existing PQ issues within the transmission and distribution grid, placement of a UPQC within the distributed generation network is useful . As a region of integration of UPQC in dg systems, analysis has been done on the subsequent 2 techniques- DC Linked and Separated DG UPQC systems.

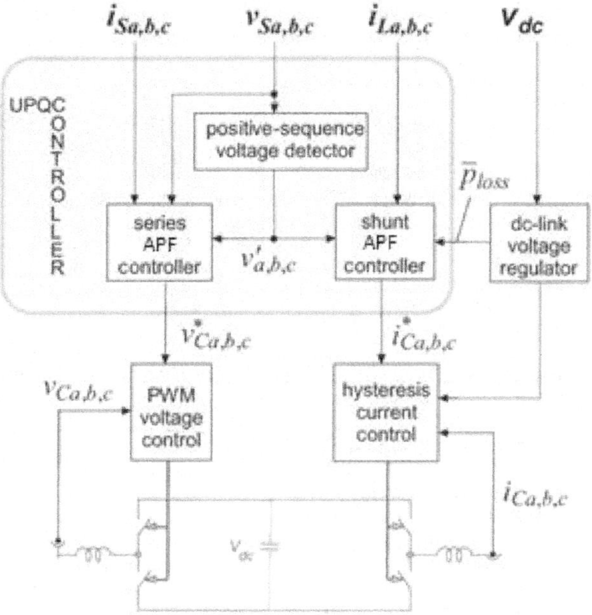

Figure 1.13: Functional block diagram of a UPQC controller

A. (DG UPQC) DC Linked

A structure has been planned in [4-7], as shown in Figure five, wherever dg sources are connected to a DC link within the UPQC as an energy source. This configuration works each in interconnected and islanded mode (shown in Figure 6). In Interconnected mode, dg provides power to the supply and loads whereas in islanded mode dg (within its power rating) provides the ability to the load solely. additionally, UPQC has the flexibility to inject power using dg to sensitive loads throughout supply voltage interruption. The advantage of this technique is voltage interruption compensation and active power injection to the grid additionally to the opposite traditional UPQC talents. The system's practicality is also compromised if the dg resources aren't comfortable throughout the voltage interruption conditions. Economical operation of the system may also be achieved by correct dominant of the active power transfer between the availability and dg supply through a series APF [7]. The planned system may also cut back the investment price by nearly one fifth if the UPQC and dg are used one by one [8].

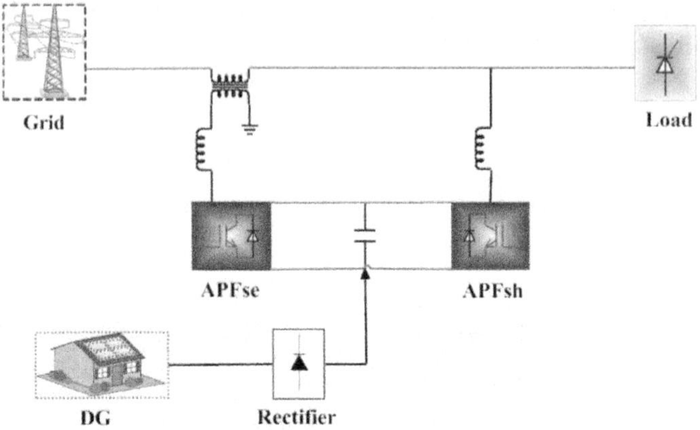

Figure 1.14: UPQC with DG connected to the DC link

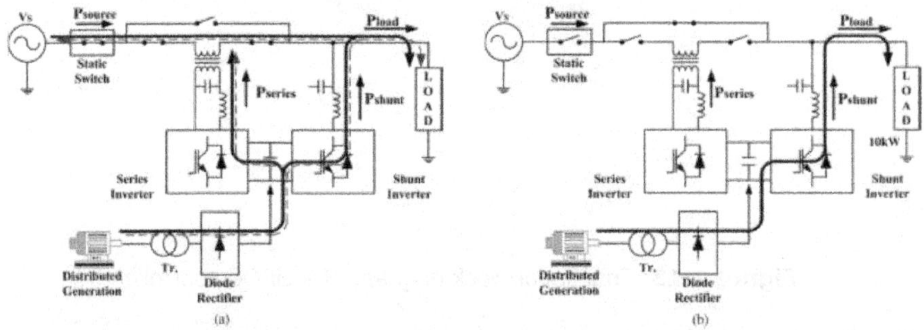

Figure 1.15: (DG UPQC) DC linked System operation concept (a) Interconnected mode; (b) Islanding mode [4]

B. (UPQC DG)$_{\text{Separated}}$

A typical application of a UPQC could be to beat the grid integration issues of the dg, like the fixed speed induction generator (FSIG) as investigated in [11] and shown in Figure seven. The FSIG fails to stay connected to the grid within the event of a grid voltage dip or line fault because of excessive reactive power demand. The call in voltage creates over-speeding of the turbine, that causes a protection trip. With the help of the UPQC, this fault-ride-through capability is achieved, that greatly enhances system stability. Results show (Figure 8) that the UPQC is one amongst the simplest devices for the mixing of wind energy system to the grid. within the case of a wind park connected to a weak grid, UPQC also can be placed at the PCC to beat voltage regulation issues [12]. In these separated systems, the series APF of the UPQC is placed close to the weight unit aspect to conduct the voltage regulation by injecting the voltage in part with PCC voltage. this sort of UPQC is stated as left shunt UPQC [13]. supported the analysis study, additionally to the conventional practicality of UPQC, a number of the opposite benefits and drawbacks are known for the techniques that are given in Table three.

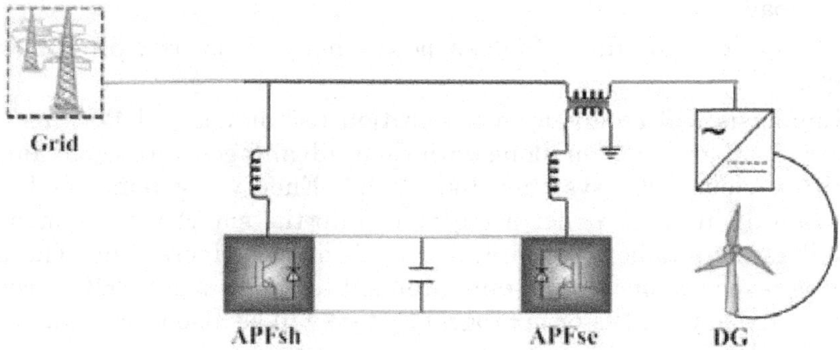

Figure 1.16: Grid connected wind energy system with UPQC

PROPOSED MODEL

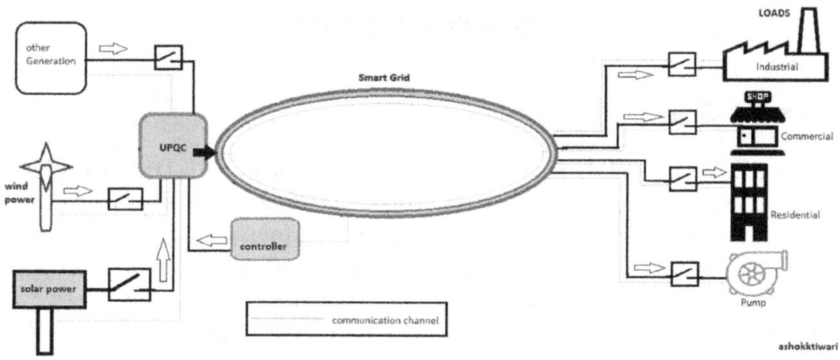

Figure 1.17: Proposed Model

1.10 Proposed Research Methodology

The MATLAB Simulation will be made for the proposed model which includes:

- Simulation of UPQC using IGBT (FACT-PE)
- Controller, pulse generators
- Solar

- Wind
- DG
- breaker
- Transformer
- line
- load

The data of Madhya Pradesh power network is used in simulation study.

Emphasis will be given on integration techniques of UPQC in smart grid or micro-grid system along with their advantages and disadvantages. Again, various DG systems like Wind Energy Systems and Solar Photovoltaic are now penetrating more into the smart grid or microgrid as well as the numbers of non-linear loads are also increasing. Therefore, existing research on capacity enhancement techniques of UPQC to cope up with the expanding DG or microgrid system will be also reviewed.

*****

CHAPTER - 2
LITERATURE REVIEW ON SMART GRID AND POWER QUALITY

2.1 Smart Grid

Fang Z. Peng [1] Transmission and distribution (T&D) networks are a vital a part of the power grid. As moving towards a smart grid, it's essential to modernize the T&D networks and build it Smart-grid ready system. The anticipation of flexible ac transmission systems (FACTS) has been popular for 3 decades. speedy advancements in power electronics technology within the past decades have directed to a replacement generation of FACTS devices. the trendy FACTS technology helps the transition of transmission networks to smart. With increasing penetration of distributed generation, the distribution network is seeing unprecedented variation in terms of its basic operation and control, integration from renewable energy to microgrid, from active control of power quality, voltage/var and frequency to self healing and islanding operation. As a key component smart grid at the distribution level, we tend to summarize this effort as an idea of resilient AC distribution systems (RACDS). The ideas of each FACTS and RACDS for a smart grid are introduced in this paper. Completely different configurations, key advantages, operational principles and world-wide installations of FACTS and RACDS devices are conferred thoroughly. The continuing and future direction of R&D resulting in newer generations of FACTS and RACDS are mentioned in the paper.

Bimal K. Bose [2] The paper provides application of power electronics in smart grid and renewable energy systems. The argument within the paper can embody trendy power semiconductor devices and their applications in bulk energy storage system, conservation of energy, electrical vehicles, renewable energy-systems etc. the essential parts of smart grid are also appraised.

Wen-Long Chin[3] increased intelligence and automation in smart grid leads to several heterogeneous applications making the most of the internet of Things, like demand response, reliability, energy delivery, efficiency, and fault recovery. However, vulnerabilities in smart grid arise thanks to public communication infrastructure and Internet-based protocols. To agitate security threats, energy massive information ought to be thoughtfully keep and processed to extract vital data, associate degreed security and blackout warnings ought to incline in an early stage. This work provides a comprehensive tutorial and survey to spotlight analysis challenges on the aforesaid problems within the Internet-of-Things-based smart grid. we tend to demonstrate that a surreptitious and blind energy massive information attack will be launched employing a replay theme. Also, we tend to interpret an intuitive geometric viewpoint for this sort of attack. The planned attack will bypass dangerous information detection with success using either DC or AC state estimation.

Jing Jiang [4] within the smart grid, duplex data exchange among customers, operators, and management devices considerably improves the potency of energy supply and consumption. However, integration of intelligence and cyber systems into a power grid will cause serious cyber security challenges and makes the system a lot of susceptible to cyber-attacks. to handle this difficult issue, this text presents defense mechanisms to either shield the system from attackers ahead or notice the existence of knowledge injection attacks to enhance the smart grid security. specializing in signal process techniques, this text introduces an adjective theme on detection of injected dangerous information at the centre. This theme takes the ability measurements of 2 serial information assortment slots under consideration, and detects information injection attacks by observation the activity variations and state changes between the 2 time slots. The planned theme has the aptitude of adaptively detection attacks as well as each no stealthy attacks and surreptitious attacks. surreptitious attacks area unit well-tried not possible to notice using standard residual based ways, and might cause a lot of dangerous effects on power systems than no stealthy attacks. it's incontestable that the planned theme can even be used for attack classification to assist system operators rank their actions to raised shield their systems, and is so terribly valuable in sensible smart grid systems

Vincenzo Bonaiuto [5] The market of electrical vehicles is in continuous growth. once the amount of the total electrical or hybrid vehicles are vital, they may represent a remarkable more resource of electrical storage in associate degree intermittent renewable power sources situation. the main target of this paper is to explain a MATLAB Simulink model compatible for the analysis of the behaviour of a parking area of electrical vehicles used as energy storage in a very smart grid atmosphere.

M. Ayar [6] Increasing readying of data technologies and low-inertia renewable energy sources into good grid fuel the uncertainties and reveal security and transient stability issues. Enhancing the soundness margins of good grids despite the cyber and physical disturbances emerges the necessity for cyber-aware sturdy controller style. Therefore, a distributed nonlinear robust controller is planned to enhance the transient stability margins of Synchronous Generators(SG) within the presence of excessive communication delay and cyber-physical disturbances. The planned controller uses phasor activity units to receive time period measurements and actuates distributed storage systems to inject or absorb power so as to accelerate stabilization of frequency oscillations of SG following a disturbance. The communication dependency exposes time delay and cyber-security problems since latency is inherent and might be excessive throughout an attack like denial of service. additionally, uncertainties in measurements challenge the stabilization method. Hence, the planned controller is intended for strength to delay and additive disturbances. a unique time delay compensation technique is developed to inject delay-free management signal into the closed loop. Toguarantee that each one chase error signals are globally uniformly ultimately delimited, novel Lyapunov-Krasovskii functionals area unit employed in the Lyapunov-based stability analysis. The simulation results validate the practicableness of the planned management framework and strength below cyber-physical sensible limitations.

Seth T. Hamman[7] As distributed, communication-based protection systems become additional rife within the emerging smart grid, the task of critically assessing their reliability has become more and more difficult because of the quality of their underlying software system styles. This paper demonstrates that the discipline of software system model checking is applied to smart grid protection code styles to carefully assess their fault tolerance. during this paper, the SPIN model checker (SPIN) is applied to a printed wide-area backup protection system (WABPS). The WABPS was specifically architected to be extremely reliable underneath numerous types of common failure situations, as well as mechanical malfunctions, incorrect device readings, and communication failures. However, thanks to its constitutional redundancy and suburbanized peer-to-peer style, shrewd its precise fault tolerance is nontrivial. This paper shows however SPIN is applied to the WABPS's style to brute-force prove the bounds of the amount and kinds of failures that may occur whereas the system remains ready to with success perform its perform. This same technique is applicable to a good kind of good grid protection software system styles, and also the data it provides is priceless to protection engineers throughout the event of recent systems, for assessing the standard of competitor styles, and for risk management functions.

Xuan Liu [8] it's been shown that an offender will stealthily launch false data injection attacks against the state estimation while not knowing the complete topology and parameter info of the complete power network. during this paper, we have a tendency to propose a heuristic yet effective methodology to work out a possible attacking region of one line, which needs less network info. we have a tendency to use six IEEE standard systems to validate the projected offensive strategy. This paper will reveal the vulnerability of the time period topology of an influence grid and is extremely useful to develop effective protection methods against topology attacks in good grids.

Amir-Hamed Mohsenian-Rad [9] Most of the present demand-side management programs focus totally on the interactions between a utility company and its customers/users. during this paper, we have a tendency to gift an autonomous and distributed demand-side energy management system among users that takes advantage of a bidirectional data communication infrastructure that is pictured within the future good grid. we have a tendency to use theory of games and formulate an energy consumption programming game, wherever the players are the users and their methods are the daily schedules of their home appliances and loads. it's assumed that the utility company will adopt adequate valuation tariffs that differentiate the energy usage in time and level. we have a tendency to show that for a standard situation, with one utility company serving multiple customers, the worldwide best performance in terms of minimizing the energy prices is achieved at the Nash equilibrium of the developed energy consumption programming game. The projected distributed demand-side energy management strategy needs every user to easily apply its best response strategy to the present total load and tariffs within the power distribution system. The users will maintain privacy and don't got to reveal the main points on their energy consumption schedules to different users. we have a tendency to conjointly show that users can have the incentives to participate within the energy consumption programming game and subscribing to such services. Simulation results confirm that the projected approach will scale back the peak-to-average magnitude relation of the overall energy demand, the overall energy prices, additionally as every user's individual daily electricity charges

Sara Mohtashami [10] This paper presents a multiyear distribution network coming up with improvement model for operation management and distribution systems capability with significant permeation of distributed generation . The model cogitates speculation in each traditional supply network and smart grid system, under consideration of firm or non firm arrangements. A group of studies on a thirty-three kilovolt real distribution network within the U.K. has been dole out to check the model. the most objective of the studies is to judge and compare the performance of various investment approaches, i.e., progressive and strategic investment. The studies additionally demonstrate the flexibility of the model to work out the best dg association points to cut back the system value. The results of the studies are mentioned during this paper.

2.2 Power Quality

Hans First State Keulenaer [11] elucidates that the price of poor PQ is high and rising. The Poor PQ could be a downside for an outsized range of business sectors. the bulk of web sites suffered from the variations within the degree of poor PQ (over 50%). The frequency of issues prevalence enhanced sometimes monthly and infrequently weekly. the nice power quality diminishes risk, decreases the value of doing business, and thereby improves the business bottom line.

Ashish Bendre [12] presents a paper that has handy data a few failure mechanism, that impacts a large form of industrial equipments. The voltage surges are typically command accountable for inscrutable instrumentality malfunction. in depth voltage observance knowledge proposes that voltage sags arise typically than voltage surges; and therefore the current surges that support voltage sag revival are the important culprits behind the instrumentality harm. This paper couldn't clarify the important reason for a majority of inscrutable instrumentality malfunction.

Johan H. R. Enslin [13] investigated that PQ problems connected with distributed power (DP) inverters, execute in massive numbers onto eight to the identical distribution network. The paper analyzed the event of harmonic interference of huge range of those electrical converters and evaluated the network interface of various inverter topologies and management choices.

Roy Billinton [14] developed the new deregulated electrical utility surroundings to assess the customer prices related to failures in power offer and therefore the consequences related to these failures. It states that the customer interruption price on account of electrical offer failure depends on several factors, like the sort of the customer, the particular load demand at the time of the outage, the length of the outage, the time of the outage and therefore the day within which the outage happens.

Jose Luis Gutierrez [15] reports that power quality is well known as a very important issue, and there's still no agreement on its total economic impact. Indeed, there's no clear analysis method to live this impact. A Joint social unit, JWG C4.107, has been shaped between CIGRE (Electric Power Transmission Emphasis) and CIRED (Electric Power Distribution Emphasis) to develop a scientific approach to the present issue. The Joint social unit plans to develop a framework for analysis of the economic science of power quality.

Roman Targosz [16] presents the results of the study conducted by sculptor Power Quality Initiative team. The study sample includes sixty two face-to-face surveys distributed in eight European countries, that has allowed for Associate in Nursing extrapolation of the general nine wastage caused by poor PQ in EU-25. The analysis of the many associated problems like user perception of PQ issues, the causes of PQ issues, equipments affected, solutions accessible and adopted are dealt well.

S. Bhattacharyya [17] presents a comprehensive survey on power quality (PQ) connected issues, as intimate by the purchasers in numerous countries. Also, Associate in Nursing investigation is completed to seek out the technical and economical impact of poor power quality. per Bhattacharya no definite SOP is out there to grasp the assorted power quality aspects at the purpose of affiliation (POC). Therefore, it's a perplexity for the network operators, the purchasers and also the instrumentality makers, on a way to share the responsibility at the POC wherever all of them meet.

Math H. J. Bollen [18] planned that the ability quality monitors within the event of a disturbance will either save the particular voltage wave shape that contains the event or the corresponding rms. The paper proposes a technique for automatic classification of installation events exploitation rms voltage measurements.

Emmanouil Styvaktakis [19] presents a paper on skilled system that's ready to classify differing kinds of installation events to the underlying causes (i.e., events) and supply helpful data in terms of power quality. a technique for event-based classification is employed, wherever a segmentation algorithmic program is 1st applied to divide waveforms into many potential events.

Yuan Liao [20] presents a fuzzy-expert system for automatic detection and classification of power quality disturbances. the assorted styles of disturbances embody voltage sags, swells, interruptions, shift transients, impulses, flickers, harmonics, and notches. it's assumed that the analyzed waveforms are offered in sampled kind. Fourier rework and moving ridge analysis are utilised to get distinctive options for the waveforms. A fuzzy-expert system is meant for analyzing the sort of the disturbance.

Mamun Bin Ibne Reaz [21] explicit that the bulk of power quality disturbances are non-stationary and passing. The detection and classification of power quality disturbances have tried to be terribly difficult. The construct of distinct moving ridge rework for feature extraction of power disturbance signal combined with artificial neural network, and symbolic logic incorporated as a robust tool for sleuthing and classifying power quality issues. This paper depicts a special kind of univariate at random optimized neural network combined with distinct moving ridge rework and symbolic logic to own a high degree of accuracy in classifying power quality disturbance. However, the neural network needs longer for coaching.

Saroj K. Meher [22] a brand new fuzzy supervised classification methodology supported aggregation operator is projected within the gift article. The projected classifier aggregates the data extracted by exploring feature-wise degree of happiness to categories. It uses a π-type membership perform and Mean (average) aggregation reasoning rule (operator). The effectiveness of the projected classifier is verified with eleven four benchmark information sets together with a time period money domain information, numerous performance measures are used for quantitative analysis of the classifier.

A. Mansoor [23] investigated the additive harmonic current characteristics of an outsized range of single phase loads. an entire analytical model for the foremost common load kind is derived. From this model the impact of interaction because of shared supply resistivity and variations in circuit parameters, on individual and additive current harmonics are derived.

W. macintosh Grady [24] explained that the single-phase power electronic primarily based loads create no drawback to the ability systems. In total, however, they need the potential to boost harmonic voltages and currents to intolerably high levels. W. macintosh projected the rules for modelling Infobahn harmonics currents created by these masses. the 2 key factors wont to verify the rules are: one. Harmonic current cancellation because of point diversity, and 2. Attenuation because of system resistivity and also the corresponding voltage distortion that tend to scale back Infobahn.

A. J. Collin [25] reported that the potential damaging effects of high influx currents occurring once power quality disturbances on switch-mode dc power provides (SMPS's), however has not been dealt very well. This work presents a brand new analytical model for the correct illustration and calculation of influx currents of low-power SMPS devices.

Rory Dwyer Afi-oz K. Khan [26] elucidates that the compact fluorescent lamps (CFLs) have the potential to extend the general harmonic distortion levels on distribution systems. The measurements of the harmonic characteristics of various varieties of CFLs were bestowed.

Daut [27] mentioned a very important thought, i.e. evaluating the impact of harmonics and their effects on power grid part and loads. Transformers are major parts in power systems. The enhanced losses because of harmonic distortion will cause excessive power losses and therefore abnormal temperature rise. This study presents the consequences of harmonic distortion of load current and voltage on single-phase electrical device losses; utilizing the on-line measurements (monitoring) methodology of the first aspect. The transformer is loaded with an SCR speed controller that consists of a single-phase full-bridge rectifier; SCR and a resistant load to get numerous total harmonic current and voltage distortions (THDi and THDv) conditions.

M. Erhan Balci [28] expressed a good concern for modelling of power electronic devices and their behaviours on power industries and customers. one in all the wide proliferated devices is that the triac controlled ac choppers, although these varieties of devices have necessary application areas, their poor power issue is that the main obstacle for the massive scale implementation in power systems. The results of those analysis show that the distortion of voltage improves the power factor, thirteen that is contradicting with earlier findings, for the high physical phenomenon angles of triac controlled ac chopper.

Juan C. Gomez [29] experimental study and a few calculations on induction motor behaviour were applied. the consequences of short interruptions and voltage sags were investigated. a regular three- phase squirrel-cage motor of 5.5 kW, 1500 r/min, and 380 V was used. The incidence of induction motor, changes the voltage sag wave and period. Besides, on-sag and post-sag currents will reach levels above the direct begin values. The Post-sag over current length will last quite double the traditional begin fundamental quantity, having specific energies within the same order of magnitude.

Math H.J. Bollen [30] reports that the voltage dips at the terminals of sensitive instrumentality are usually because of the faults that occur at a way higher voltage level. although the load current is small compared to the fault current, the changes in load current throughout and once the fault still create that the voltage at the instrumentality terminals is completely different from the voltage at the faulted voltage level. This paper mentioned the amendment within the form and magnitude of the voltage dip throughout its transmission from the faulted voltage level right down to the instrumentality terminals. each balanced and unbalanced voltage dips are mentioned by presenting theoretical issues and measurements.

*****

CHAPTER - 3
MODELLING AND SIMULATION

3.1 Modelling of Smart Grid

The proposed model of smart grid consists of existing 33/11 KV 2x 10 MVA substation, the grid incorporate renewable energy sources such as solar power plant and wind energy system. The transmission Line modeled as pi-section with appropriate resistance, inductance and capacitance suitable for ACSR Dog conductor. The Line length of 11 KV radial distribution Line is considered as 10 KM.

On the load side three different loads such as 11 KV industrial Load and on L.T. side (400V) agriculture load and domestic load are connected through the circuit Breakers of suitable ratings. An 11 KV Three phase fault is connected which is capable of providing single Line to ground, double Line, double Line to ground, three phase and three phase to ground fault.

The power sources are capable to generate third and fifth harmonies so that the effect on power quality can be accessed.

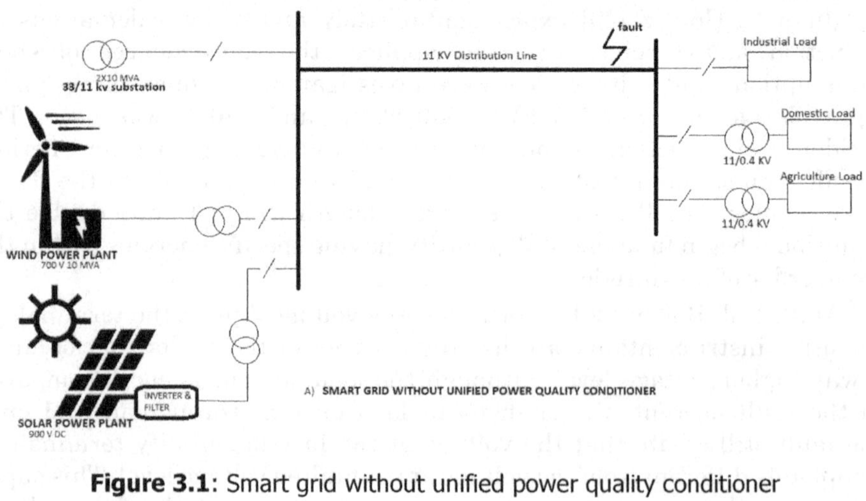

Figure 3.1: Smart grid without unified power quality conditioner

In another model of smart grid the unified power quality conditioner is connected across 11 KV Line. The sending end is connected to UPQC through series transformers and receiving end is connected as short transformer. The UPQC is modelled with series controller and short controller circuits incorporating gate triggering based on voltage & current inputs from 11 KV Line. The Capacitor of UPQC is charged with suitable DC Voltage.

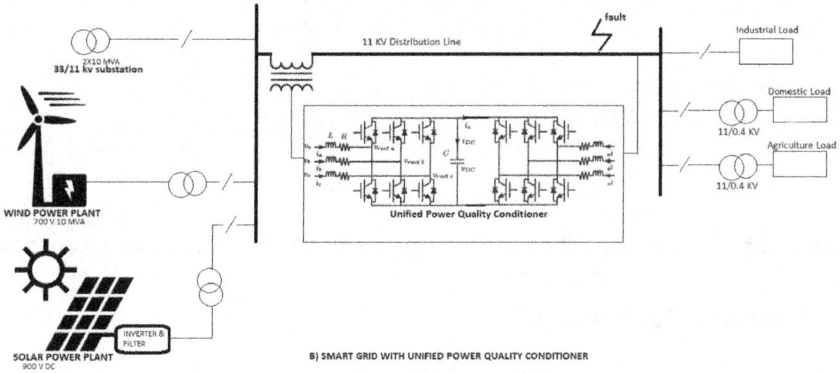

Figure 3.2: Smart grid with unified power quality conditioner

3.2 MATLAB Simulation

Both models have been simulated in MATLAB

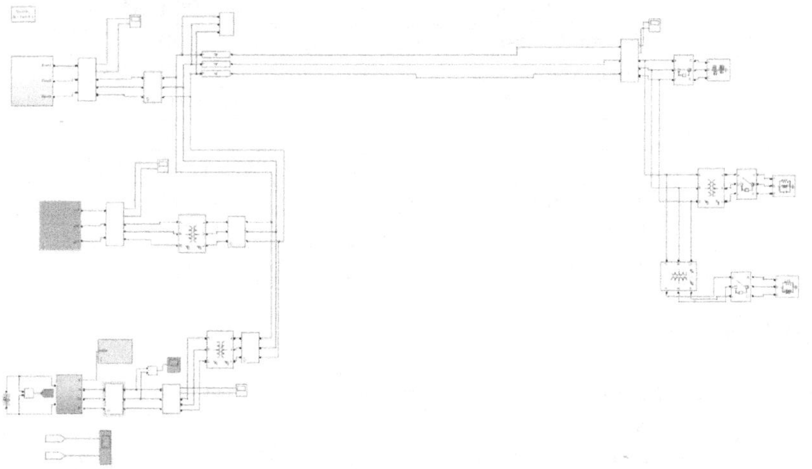

Figure 3.3: MATLAB Simulation of Smart grid without unified power quality

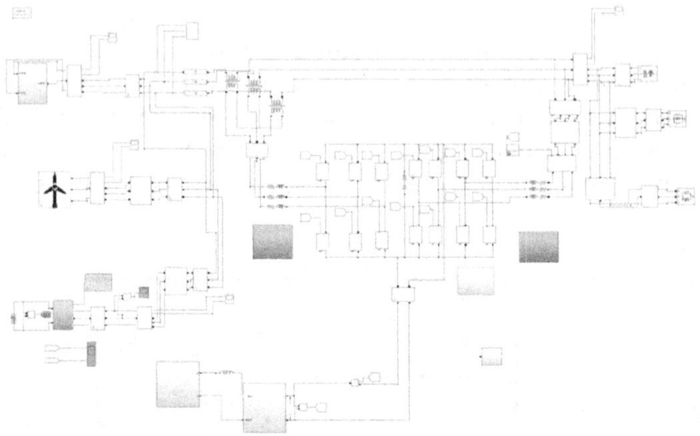

Figure 3.4: MATLAB Simulation of Smart grid with unified power quality conditioner

3.3 Network Parameters

Table 3.1: Network Parameter

	COMPONENT	PARAMETER
1	33/11 KV Conventional Grid Substation	2X10 MVA
2	Wind Turbine-Generator	700 V PH-PH rms, 5 KW X 100
	Wind Power Transformer	700/11000 V, 20 MVA
3	Solar PV	960 V DC
	Inverter Filter	960 AC
	Solar Power Transformer	960/11000 V, 1 MVA
4	Transmission Line	R/M=0.01273
		L/M=0.9337e-3
		C/M=12.74e-9
		Length 10 KM
5	Industrial Load	6 MVA
	Domestic Load	2 MVA
	Agriculture Load	1 MVA

3.4 Objectives

- Modelling of smart grid with and without UPQC through MATLAB/ SIMULINK
- Measurement of fault current and Voltage on smart grid under different fault conditions

- Calculation of energy savings with UPQC
- Observing the effect of harmonics on smart grid with UPQC

*****

CHAPTER - 4
RESULTS AND CONCLUSION

4.1 Grid Voltage and Currents Measurement

	Without UPQC						WITH UPQC					
	Grid Voltage (V)			Grid Current (A)			Grid Voltage (V)			Grid Current (A)		
	R	Y	B	R	Y	B	R	Y	B	R	Y	B
NORMAL	11000	11000	11000	618	621	619	11000	11000	11000	450	450	450
R-G FAULT	0	11000	11000	6321	700	700	0	11000	11000	4088	551	549
R-Y FAULT	0	0	11000	5516	5534	700	0	0	11000	4114	4161	700
R-Y-G FAULT	0	0	11000	6105	6119	700	0	0	11000	4211	4189	700
R-Y-B FAULT	0	0	0	6022	6112	6210	0	0	0	4501	4488	4490
R-Y-B-G FAULT	0	0	0	6266	6401	6328	0	0	0	4717	4754	4737

4.2 Waveforms

4.2.1 Normal Grid Voltages and Current

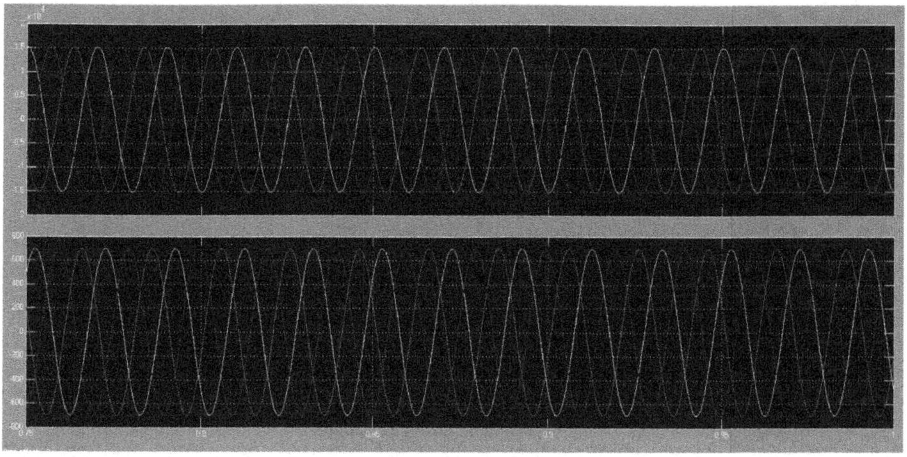

4.2.2 Grid Voltages and Current – R-phase to Earth Fault

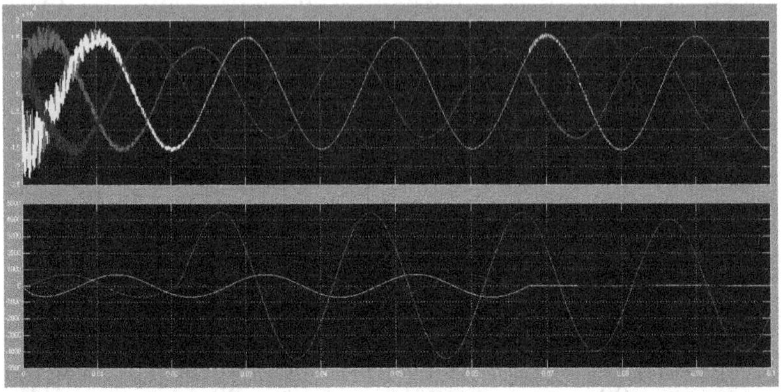

4.2.3 Grid Voltages and Current – R and Y Phase Fault

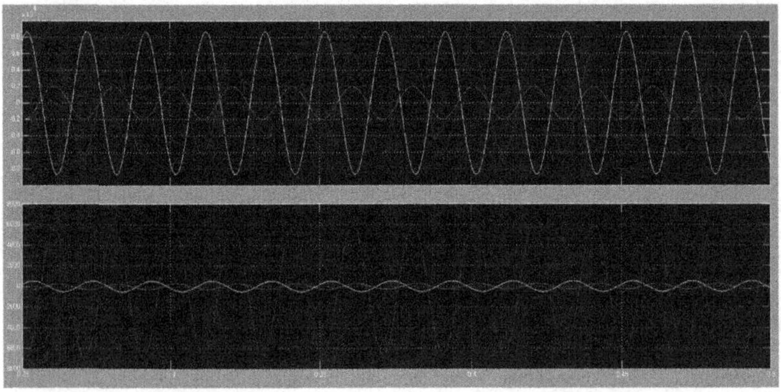

4.2.4 Grid Voltages and Current – R-Y- B Fault

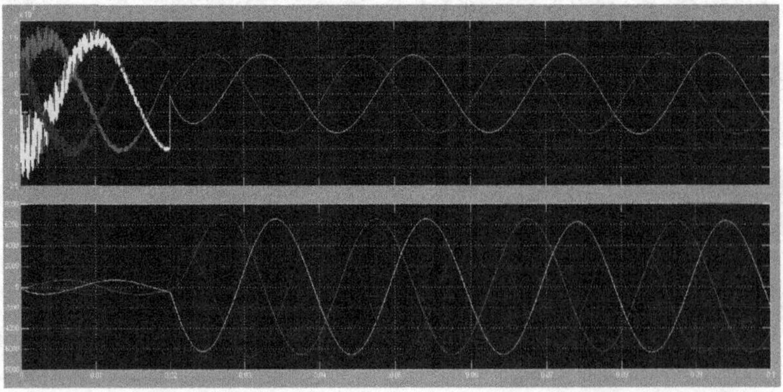

4.2.5 Grid Voltages and Current – R-Y- B to Earth Fault

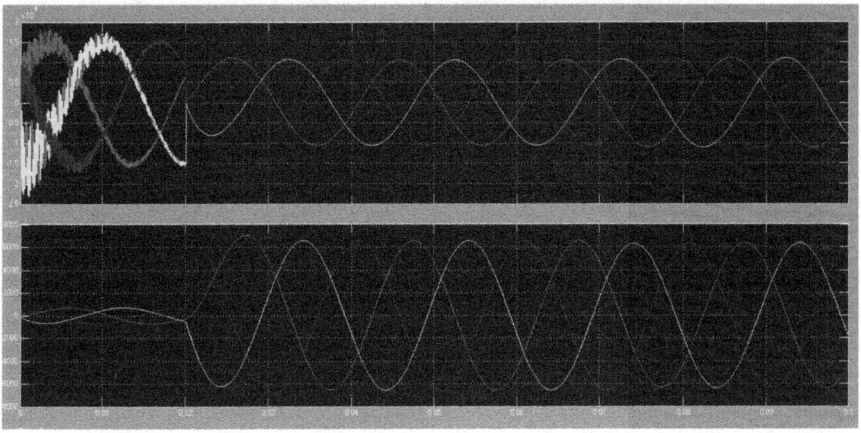

4.2.6. Solar D.C.

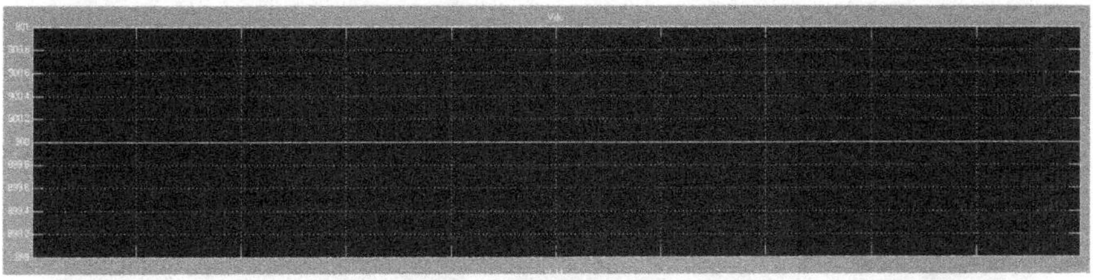

4.2.7. Solar A.C.

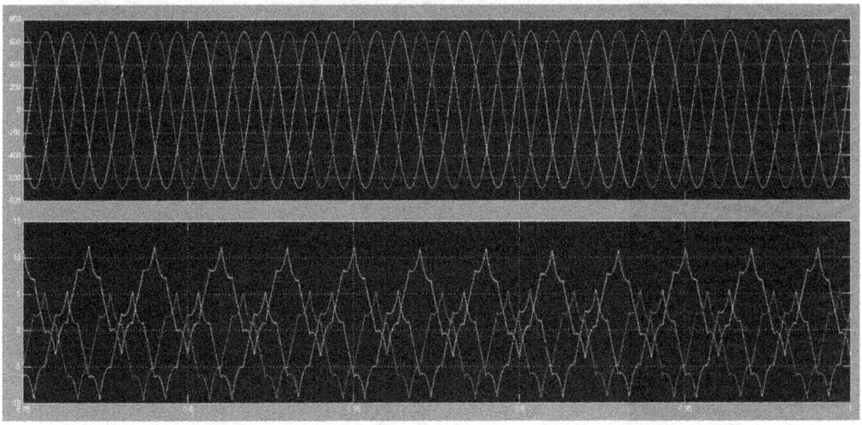

4.2.8 Solar A.C. Current after Filter

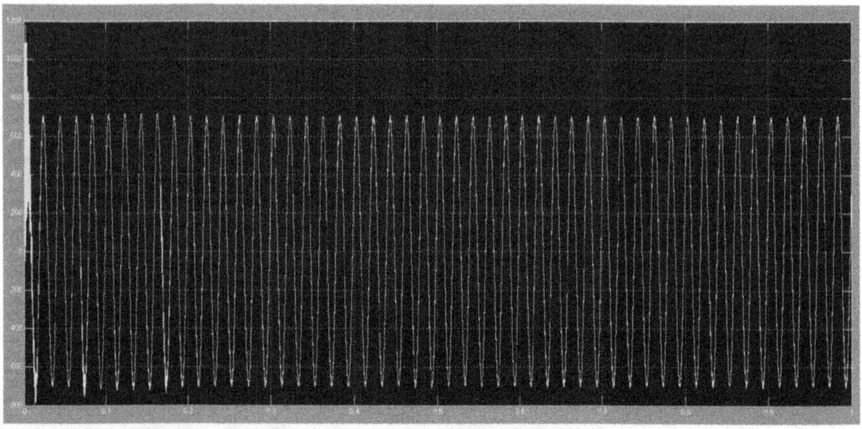

4.2.9 Wind Power

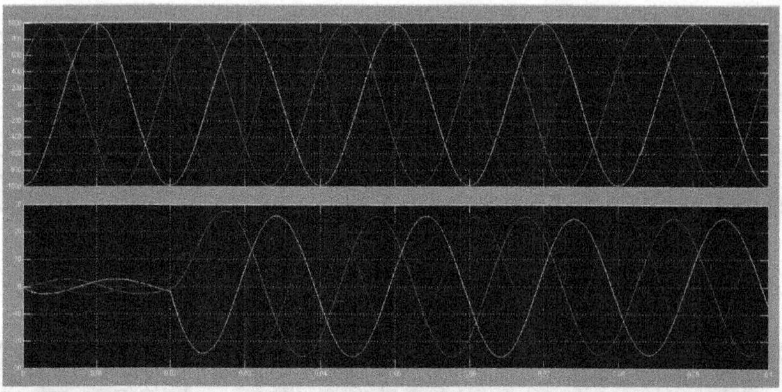

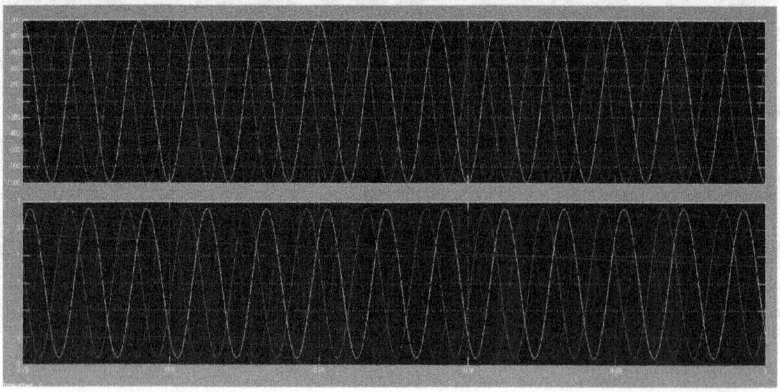

4.3 Loss Reduction

R of line 0.01273 x 10 = 0.1273 OHM

Without UPQC

V*I*t= 11000 x 619.3x0.9 x 24x30/1000 = 4414370 kwh per month

WITH UPQC

V*I*t = 11000x450 x 0.9 x 24x30/1000 = 3207600 kwh per month

Total energy saving = 1206770 kwh per month

Considering the cost of energy Rs 3.5 per kwh there is saving of Rs 4223695 per month or Rs 5 crores per annum on one 11 kv line

4.4 Harmonics Reduction

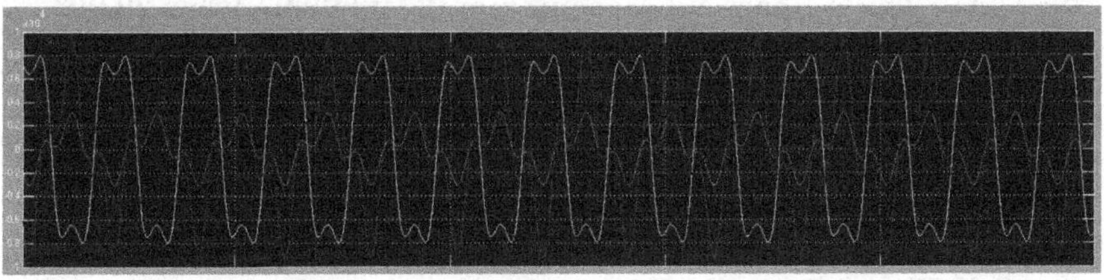

Figure 4.3: Generated harmonics waveform at source:

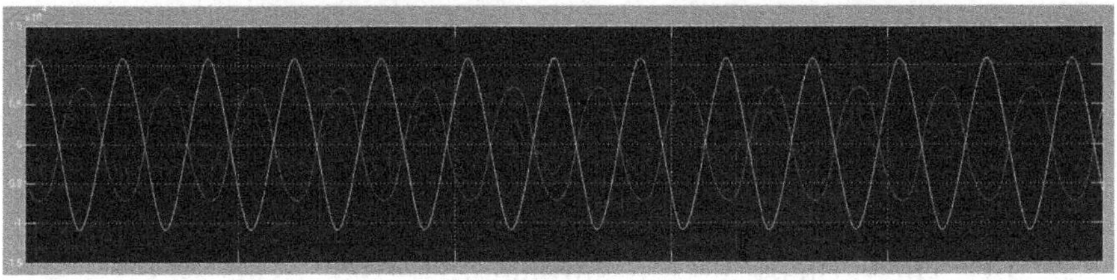

Figure 4.4: Effect of harmonics at load end without UPQC

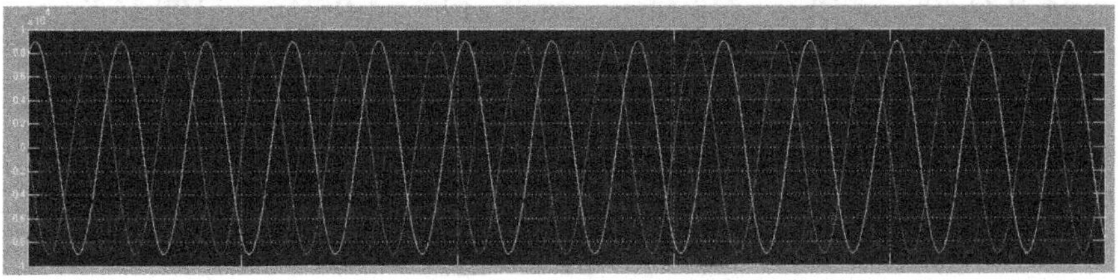

Figure 4.5: Effect of harmonics at load end with UPQC

4.5 Steady State Analysis

Without UPQC

MEASUREMENTS:

1: 'U_Vab_VSC2 ' = 488.84 Vrms 25.81°
2: 'U_Vdc ' = 0.00 Vrms 0.00°
3: 'U AB: Three-Phase V-I Measurement4 ' = 10350.23 Vrms 12.33°
4: 'U BC: Three-Phase V-I Measurement4 ' = 10350.23 Vrms -107.67°
5: 'U CA: Three-Phase V-I Measurement4 ' = 10350.23 Vrms 132.33°
6: 'U AB: Three-Phase V-I Measurement3 ' = 488.84 Vrms 25.81°
7: 'U BC: Three-Phase V-I Measurement3 ' = 488.84 Vrms -94.19°
8: 'U CA: Three-Phase V-I Measurement3 ' = 488.84 Vrms 145.81°
9: 'U AB: Three-Phase V-I Measurement2 ' = 699.97 Vrms 29.99°
10: 'U BC: Three-Phase V-I Measurement2 ' = 699.97 Vrms -90.01°
11: 'U CA: Three-Phase V-I Measurement2 ' = 699.97 Vrms 149.99°
12: 'U AB: Three-Phase V-I Measurement1 ' = 10755.87 Vrms 25.82°
13: 'U BC: Three-Phase V-I Measurement1 ' = 10755.87 Vrms -94.18°
14: 'U CA: Three-Phase V-I Measurement1 ' = 10755.87 Vrms 145.82°
15: 'I A: Three-Phase V-I Measurement1 ' = 493.84 Arms -17.66°
16: 'I A: Three-Phase V-I Measurement2 ' = 1.95 Arms -17.83°
17: 'I A: Three-Phase V-I Measurement3 ' = 0.09 Arms -97.83°
18: 'I A: Three-Phase V-I Measurement4 ' = 493.92 Arms -17.69°
19: 'I B: Three-Phase V-I Measurement4 ' = 493.92 Arms -137.69°
20: 'I C: Three-Phase V-I Measurement4 ' = 493.92 Arms 102.31°
21: 'I B: Three-Phase V-I Measurement3 ' = 0.09 Arms 142.17°
22: 'I C: Three-Phase V-I Measurement3 ' = 0.09 Arms 22.17°
23: 'I B: Three-Phase V-I Measurement2 ' = 1.95 Arms -137.83°
24: 'I C: Three-Phase V-I Measurement2 ' = 1.95 Arms 102.17°
25: 'I B: Three-Phase V-I Measurement1 ' = 493.84 Arms -137.66°
26: 'I C: Three-Phase V-I Measurement1 ' = 493.84 Arms 102.34°

With UPQC

MEASUREMENTS:

1: 'U_Vab_VSC2 ' = 640.52 Vrms 63.25°
2: 'U_Vdc ' = 0.00 Vrms 0.00°
3: 'U_Voltage Measurement ' = 0.00 Vrms 60.34°
4: 'U AB: Three-Phase V-I Measurement5 ' = 60.88 Vrms 13.63°

5: 'U BC: Three-Phase V-I Measurement5 ' = 83.92 Vrms -161.48°

6: 'U CA: Three-Phase V-I Measurement5 ' = 23.84 Vrms 31.08°

7: 'U A: Three-Phase V-I Measurement ' = 750.08 Vrms 52.90°

8: 'U B: Three-Phase V-I Measurement ' = 2796.00 Vrms -113.64°

9: 'U C: Three-Phase V-I Measurement ' = 2073.85 Vrms 71.18°

10: 'U_100Vdc DC//DC Converter3/Chopper/Voltage Measurement1' = 0.00 Vrms 60.34°

11: 'U_100Vdc DC//DC Converter3/Voltage Measurement ' = 0.00 Vrms 60.34°

12: 'U A: Three-Phase V-I Measurement1 ' = 1524.44 Vrms 77.08°

13: 'U B: Three-Phase V-I Measurement1 ' = 6347.23 Vrms -120.03°

14: 'U C: Three-Phase V-I Measurement1 ' = 4582.48 Vrms 79.51°

15: 'U A: Three-Phase V-I Measurement2 ' = 340.31 Vrms 2.43°

16: 'U B: Three-Phase V-I Measurement2 ' = 404.10 Vrms -120.01°

17: 'U C: Three-Phase V-I Measurement2 ' = 375.75 Vrms 115.08°

18: 'U A: Three-Phase V-I Measurement3 ' = 124.91 Vrms 77.03°

19: 'U B: Three-Phase V-I Measurement3 ' = 520.06 Vrms -120.03°

20: 'U C: Three-Phase V-I Measurement3 ' = 375.47 Vrms 79.49°

21: 'U AB: Three-Phase V-I Measurement4 ' = 60.88 Vrms 13.63°

22: 'U BC: Three-Phase V-I Measurement4 ' = 83.92 Vrms -161.48°

23: 'U CA: Three-Phase V-I Measurement4 ' = 23.84 Vrms 31.08°

24: 'I_Current Measurement ' = 0.00 Arms -54.89°

25: 'I A: Three-Phase V-I Measurement ' = 0.02 Arms 52.90°

26: 'I B: Three-Phase V-I Measurement ' = 0.06 Arms -113.64°

27: 'I A: Three-Phase V-I Measurement5 ' = 0.00 Arms -6.36°

28: 'I B: Three-Phase V-I Measurement5 ' = 0.00 Arms 156.65°

29: 'I C: Three-Phase V-I Measurement5 ' = 0.00 Arms -3.87°

30: 'I C: Three-Phase V-I Measurement ' = 0.04 Arms 71.18°

31: 'I_100Vdc DC//DC Converter3/Chopper/Current Measurement ' = 0.00 Arms 0.00°

32: 'I A: Three-Phase V-I Measurement1 ' = 133.85 Arms 77.02°

33: 'I B: Three-Phase V-I Measurement1 ' = 3.70 Arms -164.74°

34: 'I C: Three-Phase V-I Measurement1 ' = 4265.06 Arms 77.51°

35: 'I A: Three-Phase V-I Measurement2 ' = 2092.80 Arms -102.69°

36: 'I B: Three-Phase V-I Measurement2 ' = 2.18 Arms -165.10°

37: 'I C: Three-Phase V-I Measurement2 ' = 1396.74 Arms 77.26°

38: 'I A: Three-Phase V-I Measurement3 ' = 0.86 Arms -37.56°

39: 'I B: Three-Phase V-I Measurement3 ' = 3.19 Arms 155.90°
40: 'I C: Three-Phase V-I Measurement3 ' = 2.36 Arms -19.27°
41: 'I A: Three-Phase V-I Measurement4 ' = 0.98 Arms 27.33°
42: 'I B: Three-Phase V-I Measurement4 ' = 4.08 Arms -169.66°
43: 'I C: Three-Phase V-I Measurement4 ' = 2.95 Arms 29.83°

4.6 FFT Analysis

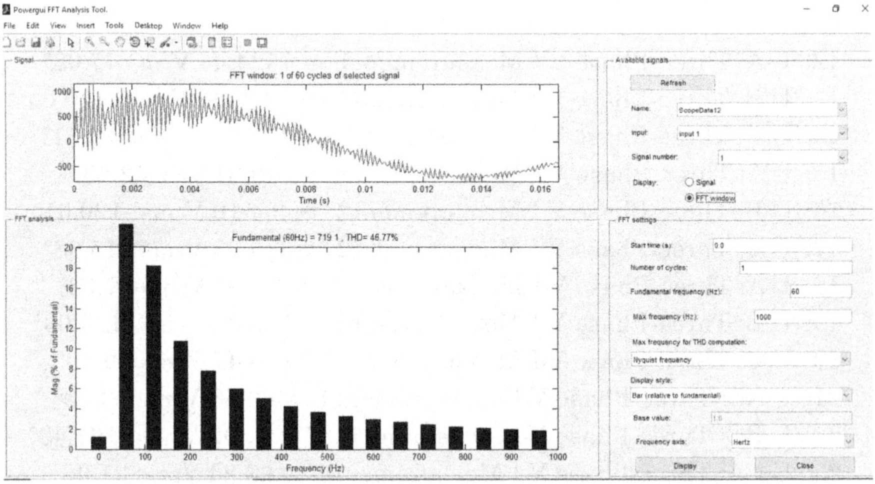

Figure 4.6: FFT Analysis without UPQC

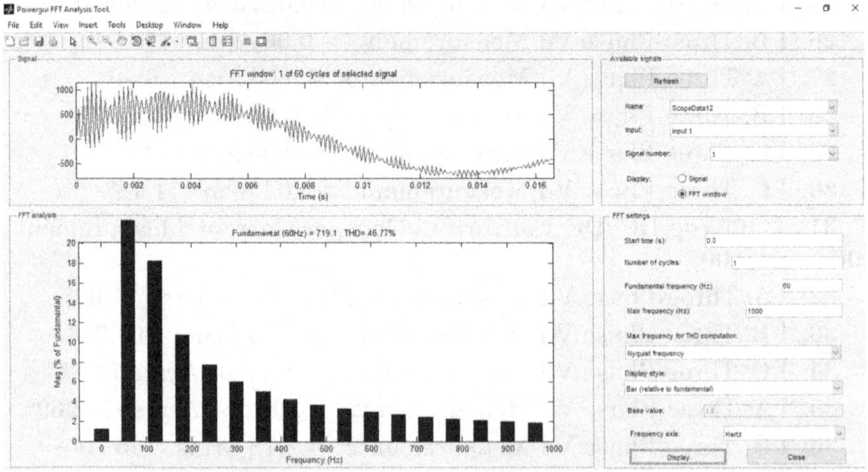

Figure 4.7: FFT Analysis with UPQC

4.8 Limitation

The study consists of 11 KV grid only. Suitable simulation can be made with other Voltage grids .

4.9 Scope for Future Work

Automatic switching and network islanding may also be incorporated in addition to UPQC

4.10 Conclusion

- It is found that using UPQS in smart grid fault current reduces to certain extent.
- It is found that load current also reduces using UPQC
- It is observed that there is reduction in energy Loss using UPQC
- Financial saving also observed up to much extent
- It is also observed that due to reduction of harmonics the power quality is enhanced.

*****

SECTION - B "MITIGATION OF VOLTAGE SAG WITH TRANSIENT AND STEADY-STATE CONTROL USING DYNAMIC VOLTAGE RESTORER"

Abstract

Due to complexity of power system combined with other factors such as increasing susceptibility of equipment, power quality is apt to waver. With electricity demand growing, low power quality is on the rise & becoming notoriously difficult to remedy [1]. Distribution system needs to be protected against voltage sags, dips & swells that adversely affect the reliability & quality of power supply at the utility end. The Dynamic voltage restorer (DVR), which has been utilized in optimized way so as to improve performance, has been put under new technique of sag detection. Various control strategies have been developed to mitigate the voltage sag/swell and the mitigation of unbalanced voltage dips. The applications of Fuzzy logic controller have taken new dimension in various fields. In this thesis, full panorama of power quality disturbances, with background theory and guidelines on measurement procedures & problem solving techniques are presented. The essentials of control scheme with immediate voltage generation to regulate the unbalance voltage phase in three phase system and a tested method to improve the reliability within the distribution system is presented. The 11kV system is controlled by linear & non-linear techniques and their performance levels are compared. The simulation results of the recommended system states the better reliability & compatibility of non-linear techniques with non-linear loads than the system with linear techniques with non-linear loads. The capability of DVR is demonstrated using MATLAB/SIMULINK simulation models. From the obtained results, we have considered the feasibility & practicability of the approach for the linear and non-linear load under voltage sag condition

Literature Review

This chapter included the study about power quality, voltage sag characteristic and the most important is a study about the main part of the project, DVR system. Thus, the formula and important theory were stated in this chapter for a better understanding during Modelling in:

1. This paper presents the Modelling and simulation of a dynamic voltage restorer as a voltage sag mitigation device in electrical power distribution networks. The dynamic voltage restorer with its excellent dynamic capabilities, when installed between the supply

and a critical load feeder, can compensate for voltage sags/swells, restoring line voltage to its nominal value within few milliseconds and hence avoiding any power disruption to the load. In this paper the technical aspect feasibility related to the use dynamic voltage restorer (DVR) of traditional DC storage systems are evaluated

2. The Dynamic Voltage Restorer (DVR) is fast, flexible and efficient solution to voltage sag problem. The DVR is a power electronic based device that provides three-phase controllable voltage source, whose voltage vector (magnitude and angle) adds to the source voltage during sag event, to restore the load voltage to pre-sag conditions. The DVR is designed for protecting the whole plant with loads in the range of some MVA. The DVR can restore the load voltage within few milliseconds.
3. The use of sensitive electronic equipment has increased now a days which has lead to power quality problems. The various power quality disturbances are transients, interruptions, voltage sag, voltage swell, voltage collapse, harmonics etc. To solve these power quality problems various custom power devices are used. Dynamic voltage restorer (DVR) is a custom power device used for the Compensation of voltage sag and swell. In this paper an overview of DVR, its components, functions, compensating strategies and control methods are reviewed in detail and the compensating strategies are compared.
4. Power qualities is one of major concerns in the present era. It has become important, especially, with the introduction of sophisticated devices, whose performance is very sensitive to the quality of power supply. Power quality problem is an occurrence manifested as a nonstandard voltage, current or frequency that results in a failure of end use equipments. One of the major problems dealt here is the voltage sag. To solve this problem, custom power devices are used. One of those devices is the Dynamic Voltage Restorer (DVR), which is the most efficient and effective modern custom power device used in power distribution network
5. Dynamic voltage restorers (DVR) can provide the most commercial solution to mitigation voltage sag by injecting voltage as well as power into the system. The mitigation capability of these devices is mainly influenced by the maximum load; power factor and maximum voltage dip to be compensated. Maximum load on a feeder is an important task for DVR system operation and appropriate desired voltage sag compensation. This paper is intended to assimilate the amount of DC energy storage depends on the installed load. It is available in a convenient manner for DVR power circui

6. Power qualities is one of major concerns in the present. It has become important, especially with the introduction of sophisticated devices, whose performance is very sensitive to the quality of power supply. Power quality problem is an occurrence manifested as a nonstandard voltage, current frequency that results in a failure of end use equipments. To solve this problem, custom power devices are used. One of those devices is the dynamic voltage restorer (DVR).The dynamic voltage restorer is one of the modern devices used in distribution systems to protect consumers against sudden changes in voltage amplitude
7. Dynamic voltage restorers (DVR) can provide the most commercial solution to mitigate voltage sag by injecting voltage as well as power into the system. This paper describes the effectiveness of using dynamic voltage restorers (DVRs) in order to mitigate voltage sags in power distribution systems at critical loads. The DVR is a power electronic based device that provides three-phase controllable voltage source, whose voltage vector (magnitude and angle) adds to the source voltage during sag event, to restore the load voltage to pre-sag conditions. The DVR can restore the load voltage within few milliseconds.
8. The problem of voltage sags and its severe impact on sensitive loads is well known. To solve this problem, The DVR is a modern and important custom power device for compensation voltage sags in power distribution systems. The Dynamic Voltage Restorer (DVR) is fast, flexible and efficient solution to voltage sag problem. The DVR is a series compensator used to mitigate voltage sags and to restore load voltage to its rated value. In this paper, an overview of the DVR, its functions, configurations, components, operating modes
9. Power qualities is one of the major concerns in the era of power system. Power quality problem occurred due to nonstandard voltage, current or frequency, that result in a failure of end user equipment. To overcome this problem, Dynamic Voltage Restorer (DVR) is used, which eliminate voltage sag and swell in the distribution line, it is efficient and effective power electronic device. The size of DVR is small, cost is low and fast dynamic response to the disturbance. By injecting an appropriate voltage, the DVR restores a voltage waveform and ensures constant load voltage
10. Power qualities has become a major area of concern in present era due to the increase in modern sensitive and sophisticated loads connected to the Distribution System. One of the major problems dealt in this paper is the voltage sag which is very severe for the

industrial customers as it can cause malfunctioning of several sensitive electronic equipments. Dynamic Voltage Restorer (DVR) is a custom power device (CPD) that is connected in series with the network to improve voltage disturbance in the electrical system

Scope of Work

Power quality gains its importance with the introduction of sophisticated electrical gadgets. Nowadays, non-linear loads cause the distortion of sinusoidal waveform which adversely affect the power quality performance. Switching of heavy loads, capacitors, and transformers and unbalanced load on a three phase system are some of the sources that contribute to voltage sag. Due to these voltage sags, the performance and the life of the equipments deteriorate considerably. This calls for the introduction & usage of Custom Power devices (CPD) with philosophy of improvement of the power quality. As per the literature review, the DVR provides excellent voltage regulation capabilities in the influence of various power quality problems. The objective of the proposed dissertation is to promise power quality & reliability in the distribution network with the simulation of various control strategies of DVR .The three control schemes namely PI Fuzzy PI-Fuzzy have been compared on account of the amount of compensation being injected into the system under voltage sag condition for linear & nonlinear loads. The three controllers provide almost equivalent compensation for linear loads but the difference in compensation occurs during non-linear loads. The capability of DVR control schemes is demonstrated using MATLAB/SIMULINK simulations. The Simulink models have been developed for the distribution networks with linear and non-linear loads. The effectiveness of PI controller based DVR, Fuzzy controller based DVR and hybrid PI Fuzzy controller based DVR in these distribution network is investigated.

*****

CHAPTER - 1
INTRODUCTION

1. Introduction

The study in this thesis consists of the design and control of Dynamic Voltage Restorer in medium power systems. The content and aim of study is provided comprehensively Electrical power quality become an important issue because of the change in the characteristics of loads connected to power system due to development of technology and increase in electricity demand. Voltage, current, frequency deviations and waveform distortions that cause equipment failure, economical loss and several negative effects are known as power quality problems. The most severe power quality problems in electrical systems are called as voltage sag and swell. Voltage sag is known as short duration reductions in the rms voltage. Another problem, voltage swell is defined as an increase in the rms voltage. Several custom power devices such as UPS, DVR, static series compensator etc. have been improved to solve these problems. Among these several custom power devices, Dynamic Voltage Restorer (DVR) is an effective solution to solve these power quality problems.

Power quality issues and resulting problems are consequences of the increasing use of solid state switching devices, nonlinear and power electronically switched loads, electronic type loads .The advent and wide spread of high power semiconductor switches at utilization, distribution and transmission lines have non sinusoidal currents [1]. The electronic type load causes voltage distortions, harmonics and distortion. Power quality problems can cause system equipment mal function, computer data loss and memory mal function of the sensitive equipment such as computer, programmable logic devices [PLC] controls, and protection and relaying equipment [1]. Voltage sag and swell are most wide spread power quality issue affecting distribution systems, especially industries, where involved losses can reach very high values. Short and shallow voltage sag can produce dropout of a whole industry. In general, it is possible to consider voltage sag and swell as the origin of 10 to 90% power quality problems [2]. The main causes of voltage sag are faults and short circuits, lightning strokes, and inrush currents and swell can occur due to a single line-to ground fault on the system, which can also result in a temporary voltage rise on the unfaulted phases [3]. Power quality in the distribution system can be improved by using a custom power device DVR for voltage disturbances such as voltage sags, swells, harmonics, and unbalanced voltage. The function of the DVR is a protection device to protect the precision manufacturing process and sophisticate sensitive electronic equipments from the voltage fluctuation and power outages [4]. The DVR has been developed by Westinghouse for advance distribution. The DVR is able to inject a set of three single-phase voltages of an appropriate magnitude and duration in series with the supply voltage in Synchronism through injection transformer to restore the power quality. The DVR is a series conditioner based on a pulse width modulated voltage source inverter, which is generating or absorbing real or reactive power independently. Voltage sags caused by unsymmetrical line-to-line, line to ground, double-line-to-ground and symmetrical three phase faults is affected to sensitive loads, the DVR injects the independent voltages to restore and maintained sensitive to its nominal value. The injection power of the DVR with zero or minimum power for compensation purposes can be achieved by choosing an appropriate amplitude and phase angle [4] [5]. Section 2 discusses the basic configuration of DVR. The various operating modes of DVR are discussed in section 3. Section 4 presents the type of control strategies in DVR with linear and non – linear control. Section 5 discusses the compensation techniques in DVR. The control algorithm and conclusion are discussed in section 6 and 7 respectively.

1.1 DVR

The dynamic voltage restorer (DVR) is the most effective and economical custom power device applied to protect sensitive loads from voltage sags and swells. Dynamic voltage restorer is a series connected device located between sensitive load and grid in system, it both detects voltage sags/ swell problems and injects controlled voltage to system. To perform this process, a conventional DVR consists of inverter, dc-link capacitor, filter and transformer which will be extensively explained in thesis.A schematic diagram of the DVR incorporated into a distribution network is shown in Figure 1.1. VS is the source voltage, V1 is the incoming supply voltage before compensation, V2 is the load voltage after compensation, Vdvr is the series injected voltage of the DVR, and I is the line current. The restorer typically consists of an injection transformer, the secondary winding of which is connected in series with the distribution line, a voltage-source PWM bridge inverter is connected to the primary of the injection transformer and an energy storage device is connected at the dc-link of the inverter bridge. The inverter bridge is filtered in order to mitigate the switching frequency harmonics generated in the inverter. The injection of an appropriate Vdvr in the face of an up-stream voltage disturbance requires a certain amount of real and reactive power supply from the DVR. It is quite usual for the real power requirement of the DVR be provided by the energy storage device in the form of a battery, a capacitor bank, or a flywheel. The reactive power requirement is generated by the inverter.

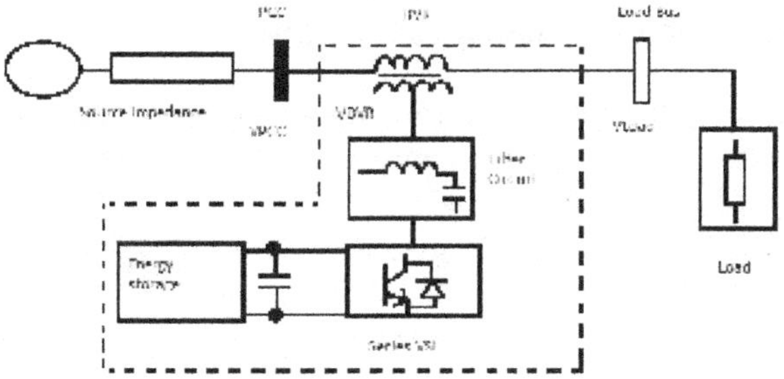

Figure 1.1: DVR

1.2 Overview

The thesis consists of the following chapters: After this first chapter, in Chapter 2 Power Quality Problems; Sources and effects of power quality problems, power quality problems and their explanation, Power Quality Standards and solution of power quality problems are explained. Chapter 3 The Fundamentals of DVR gives an overview of available DVR system, the function of converters in power circuit configuration, filter and transformer design. It also includes control methods in available literature. The power circuit design of multilevel inverter based DVR components, sag detection methods, voltage injection strategies used in proposed DVR are explained in Chapter 4 Modelling of Proposed DVR. Simulation results which consist of comparison of sag/swell detection methods, voltage injection strategies (presag, in-phase) and voltage control strategies are given in Chapter 5 Simulation Results and Case Studies. Also, this chapter includes analysis of full bridge dc-dc converter for voltage sag compensation.

1.3 Problem Statement

Nowadays, there are many quality problems such as voltage sags and swells, flicker, interruptions, harmonic distortion and other distortions to the sinusoidal waveform of the customer voltage affect the performance of the sensitive load. Power quality has now become an important component of service reliability to both utility and customer. The power quality problem has now become more critical due to industrial equipment becoming more sensitive to minor voltage variations. Industrial customers using sensitive equipment suffer a huge loss because of voltage sag. If the magnitude and duration of voltage sag exceed the sensitivity threshold of the equipment, the equipment can be damaged or become malfunctions. Based on records by TNB, 80% of power quality complaints in Malaysia were traced to be related to voltage sag [1]. Due to the increasing of new technology, a lot of devices had been created and developed for mitigation of voltage sag.

1.4 Voltage and Power Quality Issues

In an ideal world, power distribution companies should have a system in place that provides their customers with a smooth sinusoidal voltage with a fixed amplitude and frequency. However, power systems, especially at the distribution level, are composed a number of non-linear loads that negatively affect the power quality of the grid. The impact of these non-linear loads is growing [7]. The large influx of grid-tied renewable energy sources in recent years has greatly impacted power quality and total system operation. Unfortunately, the output power for most renewable energy sources is highly variable. Wind generators only produce rated power when the wind is blowing. Similarly, the efficiencies of solar PV installations are affected by the presence of clouds. The randomness associated with these energy sources results in voltage fluctuations and a decrease in power quality, both of which have profound effects on the overall system [8]. In addition to the rise of non-linear loads, there will always be random sags, swells, and surges on utility lines that affect power quality. Surges, spikes, sags, swells, harmonic distortions and momentary disruptions are examples of voltage fluctuations that can be harmful to a system. Voltage sags are the most common type of occurrence affecting power quality. They are usually caused by external factors such as storms and animals, but they can also be caused by internal factors like heavy loading and grounding problems. Voltage sags can do damage to electronic devices, as most are supposed to be operated in a specific and constant voltage range. Some of the more sensitive devices in industry include variable frequency drives, robotics, programmable logic controllers, controller power supplies, and control relays. In order to protect expensive devices like these from harmful voltage variations and maintain power quality for the electrical grid, new management strategies and devices are being implemented [9].

1.5 FACTS Controllers

FACTS controllers are static electronic controllers used to increase power quality in distribution systems with a voltage rating between 1 and 38 kV [14]. The first generation of FACTS controllers are thyristor based, whereas the second generation controllers are converter based. Generally speaking, FACTS controllers can be imcategorized into 4 different types: series controllers, shunt controllers, combined series-series controllers, and combined series shunt controllers [15]. The research portion of this thesis covers a mixture of both first and generation FACTS controllers. 2.2.1 Static VAR Compensator (SVC) Static VAR compensators (SVCs) are well known FACTS controllers that are connected in shunt with a load in order to provide voltage support and control reactive power [15]. The control of reactive power is associated with voltage support, which is to say that controlling voltage in a power system is achieved by either absorbing or generating reactive power. In an ac power system, capacitors generate reactive power and inductors absorb reactive power. This is important because SVCs are composed of static VAR generators (SVGs), which are semiconductor power circuits that use thyristors along with inductors and capacitors to produce a variable reactive output. 10 Currently, the four most common types of SVGs are the thyristor-controlled reactor (TCR), the thyristor-switched capacitor (TSC), the fixed-capacitor, thyristor-controlled reactor (FC-TCR) type VAR generator, and the thyristor-switched capacitor, thyristor controlled reactor (TSC-TCR) type VAR generator [16]. The TCR type VAR generator is

However the problem can also be mitigated by improving the immunity of equipment. The mitigation methods includes reducing the number of faults, faster fault clearing, improved network design & operation, improved end-user equipment [5].The higher index of reliability & power quality to satisfy the customer has reflected the need for the development & application of compensation systems. Compensating systems also known as the custom power devices offer a handful of protection & security to the system under observation. A custom power device is a reliable & flexible solution to the consumers regarding the power supply. They tend to absorb the various disturbances by injecting appropriate voltage, current or both into the system; thereby relieving the main source from meeting the reactive power demand of the load. Various custom power devices covering a wide range of flexible controllers & capitalize on evolution of power electronic controllers are widely used to compensate voltage sag/swells in the system [3].The custom power devices used for transmission system includes static synchronous compensator (STATCOM), static synchronous series compensator (SSSC), interline power flow conditioner (IPFC) & unified power flow conditioner(UPFC) and for distribution system are distribution static synchronous compensator(DSTATCOM), dynamic voltage restorer(DVR), active power filter(APF),unified power quality conditioner(UPQC) etc. There are various other custom power devices such as Battery Energy Storage Systems (BESS), Surge Arresters (SA), Super conducting Magnetic Energy Systems (SMES), Static Electronic Tap Changers (SETC), Solid-State Transfer Switches (SSTS) and Solid State Fault Current Limiter (SSFCL). In this work, the effectiveness of DVR to compensate load voltage is investigated. It is a series compensating device that helps in increasing the immunity of the equipment & reliability of the system by regulation of voltage in the system. The whole work is concentrated around DVR & its various control strategies. Composed of a fixed inductance and bidirectional thyristor switches. The current in the inductor can be varied by introducing a delay angle alpha to the firing pulses given to the gate of the thyristors. For maximum inductance, the alpha is set to zero so that the thyristors are always conducting. If alpha is adjusted to 180 degrees, the thyristors are always off, so the device doesn't draw any current. TCRs introduce inductive impedance to the system, meaning that the current through the reactor lags the system voltage [16]. So, this device can be adjusted to absorb a certain amount of reactive power. TSC type VAR generators are composed a capacitor, bidirectional thyristor switches, and a small reactor used to protect the device from large surge currents. The thyristors are controlled similarly to those in the TCR type VAR generators, except in this case the capacitance is dictated by the delay angle alpha. A zero degree firing angle will introduce the maximum capacitance to the system. Additionally, the current through the capacitor will lead the system voltage [16]. As the name implies, the FC-TCR type VAR generator is composed of a fixed capacitor in parallel with a TCR. It is common knowledge that ac systems are usually composed of inductive loads that create an undesired non-unity power factor. The fixed capacitor in the FC-TCR is used for power factor correction, as it is shunted near the inductive load in order to supply reactive power. However, loads are almost never fixed, 11 so the ability to absorb reactive power in addition to supplying it is very important. This is why the FC-TCR includes the parallel TCR branch. The parallel TCR allows a controllable amount of VAR to be absorbed from the system. The combination of the fixed capacitor in parallel with the TCR allows for more adjustability and better system reactive power control [16]. The TCR and TSC can be placed in parallel to form the TCR-TSC type VAR generator. The combination of the two controllable branches allows for an even more dynamic and accurate reactive power compensation [16]. The four SVG types discussed can be used in SVC controllers to produce the net reactive power needed by the load on a continuous basis. Figure 2.2 shows the three physical system devices used in SVC controllers. In addition to the TCR, the TSC and the fixed capacitor, a step-down transformer is used in an SVC to "step" the system voltage down to the 1 to 38 kV range discussed previously [16]. Although the SVC provides voltage support and reactive power control, one of the major disadvantages of the SVC is the harmonics that it produces at lower frequencies due to the switching thyristors. This in turn results in the need for heavy filtering, which takes up a lot of space and adds to the cost of the controller [17].

1.6 Thyristor-Controlled Series Capacitor (TCSC)

The thyristor-controlled series capacitor is one of the most common FACTS controllers in modern power systems. TCSCs are series compensation devices that possess the ability to quickly adjust line impedance in order to control the flow of power, improve stability, and increase the amount of transferrable power in a system. As seen in Figure 2.3, a basic TCSC is composed of a fixed capacitor in parallel with a TCR. TCSCs have four modes of operation depending on what net series reactance is needed on the line. These four operating modes are thyristor-blocked mode, thyristor-bypassed mode, and thyristor conduction mode in either the inductive or capacitive regions. 13 Starting from a zero degree firing angle, the equivalent series impedance of the TCSC is inductive. The firing angle can be increased to create a capacitive reactance, but a region near parallel resonance is prohibited. In parallel resonance, the reactance of the capacitor is equal to the reactance of the reactor, so in theory the equivalent TCSC reactance becomes infinite [18]. In terms of FACTS controllers, some consider the TCSC controller to yield the best results in terms of performance and flexibility in power distribution systems. A typical transmission line has inherent inductance, and a TCSC controller can be tuned to provide a series capacitive reactance that reduces the overall reactance of the line. In this application, the lower line reactance lowers the voltage drop across the line and increases the allowable transferable power capability of the system. So, TCSC controllers have the ability to control power flow, reduce loss, and provide voltage support. Limitations do exist, however, as TCSCs must be operated within the allowable ranges of the firing angle alpha even when the voltage is dynamic. This means that the line reactance needed from the TCSC is variable, not static, so the operating limitations of the controller must always be considered [19].

1.7 STATCOM

Another common FACTS controller is the static synchronous compensator (STATCOM). STATCOMs are shunt devices used to control reactive power flow, regulate voltage, and thus improve voltage quality without the use of reactive energy components. The most common STATCOMs are usually composed of an inverter with a capacitor on the dc side, a coupling transformer, and a control system. The reactive power output of the STATCOM is determined by the difference in magnitude between the ac system voltage and the ac voltage outputted by the voltage source converter (VSC). The amplitude of the STATCOM's output voltage is controlled by varying the magnitude of the dc voltage across the capacitor. So, the output voltage can be adjusted to make the STATCOM act like a capacitive or inductive load by either generating or absorbing reactive power, respectively. If the amplitude of the STATCOM's output voltage is adjusted to be smaller than the magnitude of the system voltage but in phase with it, then the STATCOM acts like a capacitive load [20]. A diagram of a STATCOM can be seen in Figure 2.4. STATCOMs are proven to improve power quality and provide dynamic voltage response. However, just like any other FACTS controllers in industry, STATCOMs have a few downsides. For example, the VSCs used in STATCOM controllers create large switching losses that decrease its efficiency. Additionally, the performance of a STATCOM device decreases significantly with the addition of negative sequence components in the system.

This is a problem because more control circuitry must be implemented in this case, which adds to the complexity and price of the controller [21].

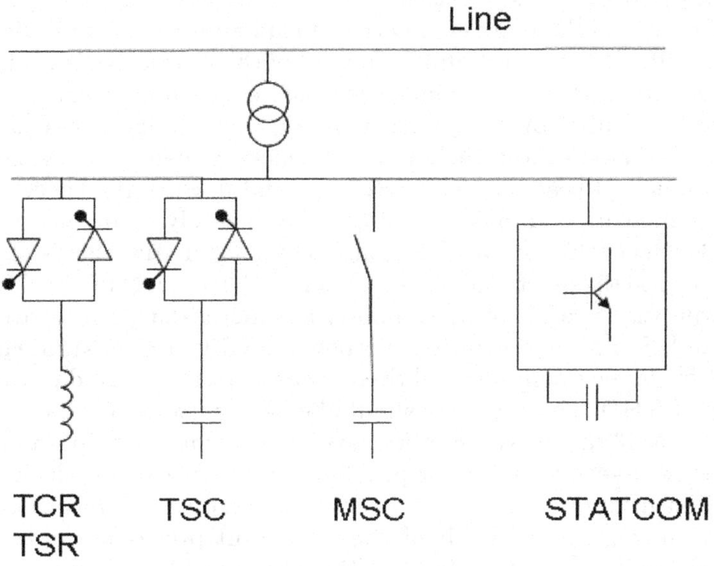

Figure 1.7: FACTS controller

*****

CHAPTER - 2
POWER QUALITY

2. Introduction

The term "Power Quality" is defined as "Set of parameters defining the properties of power quality as delivered to the user in normal operating conditions in terms of continuity of supply and characteristics of voltage (symmetry, frequency, magnitude, waveform) in IEC. In IEEE Std. 1100-1 999, "Power Quality" is defined as "The concept of powering and grounding electronic equipment in a manner that is suitable to the operation of that equipment in a manner that is suitable to the operation of that equipment and compatible with the premise wiring system and other connected equipment (Ise et al., 2000). Power Quality just meant the ability of utilities to provide electric power without interruption. However, in recent years, power quality becomes an important concern to customers as well as utilities and facilities. Customers require higher quality of power than ever before due to the increase in critical load and electronic device. Power quality is different from reliability in view of the duration of events it deals with. It treats very short events with a few cycles or seconds duration. New power quality problems such as sag, swell, harmonic distortion, unbalance, transient, and flicker may impact on customer devices, causes malfunctions and cost on lost production and downtime. These problems should be measured and assessed more accurately than before (Won et al., 2003). Recently, an increased number of sensitive loads have been integrated into the electrical power systems. Consequently, the demand for high power quality and voltage stability has increased significantly (Meyer et al., 2008). Some basic criterions for power quality are constant rms value, constant frequency, symmetrical three-phases, pure sinusoidal wave shape and limited THD. These values should be kept between limits determined by standards if the power quality level is considered to be high. Power quality covers several types of problems of electrical supply and power system disturbances. The cost of power interruptions and disturbances can be quite high as a result of the important processes controlled and maintained by the sensitive devices (Dong et al., 2004). Sources and effects of power quality problems can be summarized

Sources and Effects of Power Quality Problems Power distribution systems, ideally, should provide their customers with an uninterrupted flow of energy at smooth sinusoidal voltage at the contracted magnitude level and frequency. However, in practice, power systems, especially the distribution systems, have numerous nonlinear loads, which significantly affect the quality of power supplies. As a result of the nonlinear loads, the purity of the waveform of supplies is lost. This ends up producing power quality problems (Jena). The distortion in the quality of supply power can be occurred because of various devices; some of the primary sources of distortion can be identified as below

- Power Electronic Devices
- IT and Office Equipments
- Arcing Devices
- Load Switching
- Large Motor Starting § Embedded Generation
- Electromagnetic Radiations and Cables
- Storm and Environment Related Causes etc.

The growth of the nonlinear loads like the devices with switching power supplies have increased the current harmonics, EMI problems, unnecessary reactive power and power losses which causes distortion, harmonics, flicker phenomena, sag and swell conditions on the line voltages and other problems (Hosseini et al., 2006).Sources and Effects of Power

Quality Problems Power distribution systems, ideally, should provide their customers with an uninterrupted flow of energy at smooth sinusoidal voltage at the contracted magnitude level and frequency. However, in practice, power systems, especially the distribution systems, have numerous nonlinear loads, which significantly affect the quality of power supplies. As a result of the nonlinear loads, the purity of the waveform of supplies is lost. This ends up producing power quality problems (Jena). The distortion in the quality of supply power can be occurred because of various devices; some of the primary sources of distortion can be identified as below:

- Power Electronic Devices
- IT and Office Equipments
- Arcing Devices § Load Switching
- Large Motor Starting § Embedded Generation
- Electromagnetic Radiations and Cables
- Storm and Environment Related Causes etc

The growth of the nonlinear loads like the devices with switching power supplies have increased the current harmonics, EMI problems, unnecessary reactive power and power losses which causes distortion, harmonics, flicker phenomena, sag and swell conditions on the line voltages and other problems

2.1 Power Quality

Standard of IEE 1100-1999 defines that power quality is "The concept of powering and grounding electronic equipment in a manner suitable to the operation of that equipment and compatible with the premise wiring system and other connected equipment" [2]. Nowadays, there are many types of power quality disturbance which a power feed can be poor in quality, and so no single figure can completely measure the quality of a power feed. For example, harmonic, voltage imbalance, interruptions, voltage swell and last but not least is voltage sags.

2.2 Definition of Voltage Sag

Voltage sags can be defined as a short reduction in voltage magnitude for durations of time, and is the most important and commonly occurring power quality issue of our days. Voltage sags also known as voltage dips. The IEEE STD 1159-1995 defines voltage sag as a decrease in root mean square voltage, (RMS) to values between 0.1 p.u to 0.9 p.u, for durations of 0.5 cycles to 1 minute. In other word, voltage sags is a short term reduction of RMS voltage [2]. So that, the full required energy is not being delivered to the load and others event was explained in Appendix A.

2.3 Power Quality Problems

The importance of power quality (PQ) has risen very considerably over the last two decades due to a marked increase in the number of equipment which is sensitive to adverse PQ environments, the disturbances introduced by nonlinear loads, and the proliferation of renewable energy sources, among others. At least 50%of all PQ disturbances are of the voltage quality type, where the interest is the study of any deviation of the voltage waveform from its ideal form. The best well-known disturbances are voltage sags and swells, harmonic and inter harmonic voltages, and, for three-phase systems, voltage imbalances (Roncero-Sanchez et al., 2009). A number of national and local surveys helped to quantify the statistical aspects of this problem. The most common disturbances and the most commonly affected equipments are illustrated in Figure 2.1

2.4 Voltage Sag

Voltage sags are now one of the most important power quality problems in the distribution system. A voltage sag is a momentary decrease in the RMS ac voltage (10%–90% of the nominal voltage), at the power frequency, of duration from 0.5 cycles to a few seconds. Voltages sags are normally caused by short-circuit faults such as a single-line-to-ground fault in the power system or by the starting up of induction motors of large rating. Voltage sags may cause the malfunction of voltage sensitive loads in factories, buildings, and hospitals (Kangarlu et al., 2010). Voltage sags can cause tripping of contactors, motor starters, relays, restarting expense of computers and shutdown of an entire production line. The sources of voltage sags are basically faults on adjacent feeders, lighting, short circuit event, start up of heavy loads, transformer energizing and motor starting (Bollen, 2001).

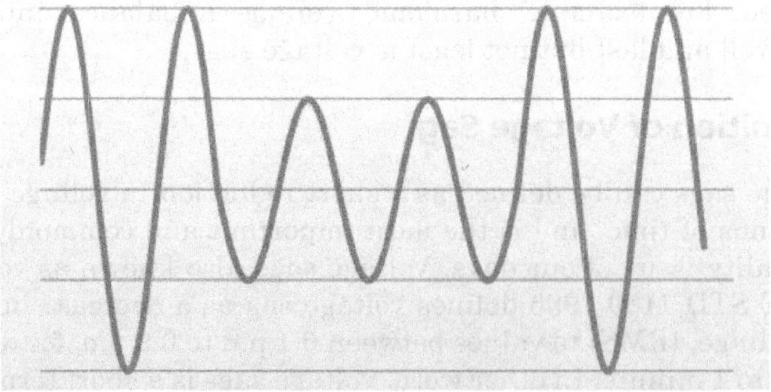

Figure 2.4: Voltage Sag

2.5 Voltage swell

Voltage Swell is defined as a short duration increasing in RMS supply with increase in voltage ranging from 1.1 pu to 1.8 pu of nominal supply. The main causes for voltage swell are switching of large capacitors or removal of heavy loads (Kangarlu et al., 2010). Voltage swells might not be as common as voltage sags, however are much more harmful and disruptive to static power converters. In fact, they may severely Voltage Swell damage or trip them, causing shutdowns of entire processes. This overall situation has become even more critical given the recent industrial trend to increase operating voltages of power converters (a practice that has pushed semiconductor devices up to their limit. A closer look to this phenomenon is, hence, required (Burgos et al., 2005)

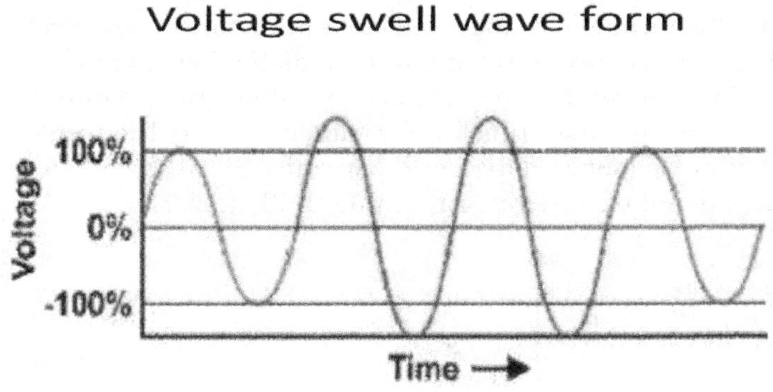

Figure 2.5: Voltage swell

2.6 Voltage Fluctuations

Voltage fluctuations are systematic variations of the voltage envelope or a series of random voltage changes. Arc furnaces are the most common cause of voltage fluctuations on the transmission and distribution system (Martinez, 1998). The voltage fluctuation is one of the major power quality problems in a weak power system, which feeds fluctuating loads, such as electric arc furnaces and arc welders. In general, the flicker components exhibit frequencies in the range 0.1 Hz to 30 Hz and are especially important due to visual irritation. Many reports indicate that a small voltage fluctuation from 0.3% to 0.5% in the frequency range of 6-10 Hz will cause visible incandescent lamp flickering and let people feel uncomfortable (Wu et al., 2006).

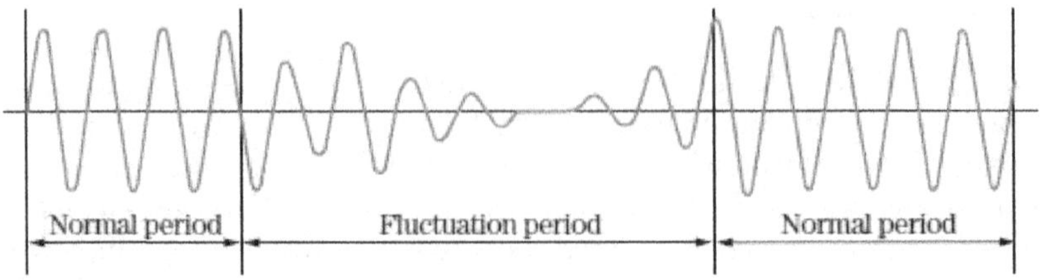

Figure 2.6: Voltage Fluctuations

2.7 Harmonics

Harmonics can be defined as the spectral components at frequencies that are integer multiples of the ac system fundamental frequency (Testa et al., 2007). The harmonic voltage and current distortion are strongly linked with each other because harmonic voltage distortion is mainly due to non-sinusoidal load currents. Current harmonic distortion requires overrating of series components like transformers and cables. As the series resistance increases with frequency, a distorted current will cause more losses than a sinusoidal current of the same rms value (Bollen, 2001).

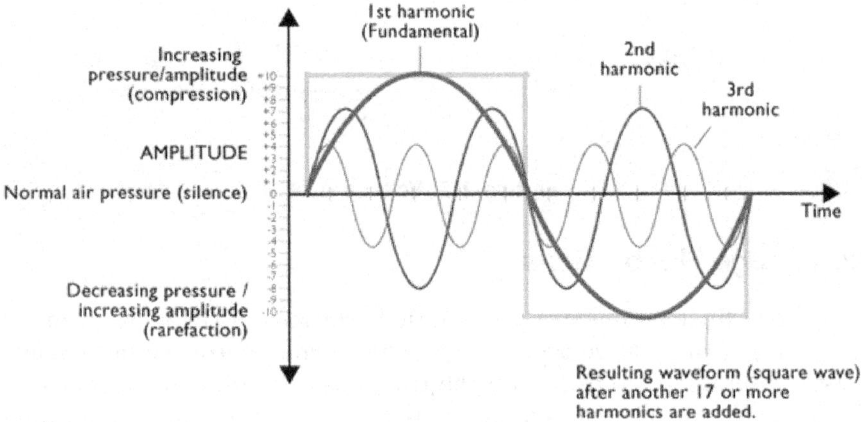

Figure 2.7: Harmonics

Interharmonics Interharmonics are spectral components at frequencies that are not integer multiples of the system fundamental frequency. Interharmonics can be observed in an increasing number of loads in addition to harmonics. The main sources of interharmonics are static frequency converters, cycloconverters, high voltage direct current (HVDC) transmission systems, induction motors, welding machines, arc furnaces, and all loads not pulsating synchronously with the fundamental power system frequency (Tayjasanant et al., 2005;Yacamini, 1996)

2.8 Transients

A transient is "that part of the change in a variable that disappears during transition from one steady state operating condition to another". Another word in common usage that is often considered synonymous with transient is "surge" (Dugan et al, 2003). Transients can be be classified into two categories: "impulsive" and "oscillatory": § Impulsive transients: Sudden, non-power frequency change in the steady state condition of the voltage, current or both § Oscillatory transients: Voltage or current whose instantaneous value changes polarity rapidly.

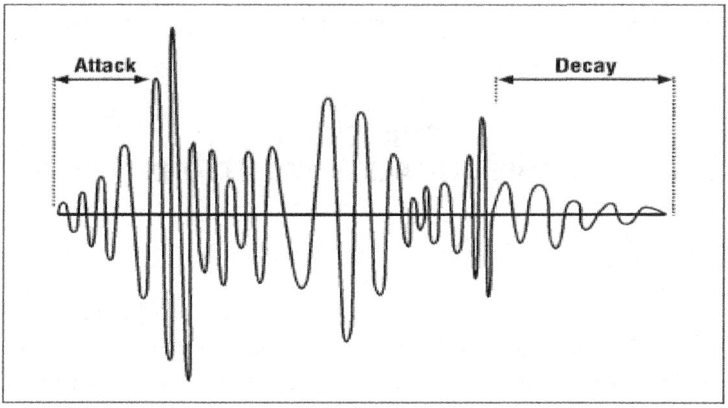

Figure 2.8: Transients

2.9 Interruption

In the European standard EN 50160 two terms are used (Nielsen et al., 2002): § Long interruptions: longer than three minutes. § Short interruptions: up to three minutes. Interruptions are typically caused by different types of faults e.g. malfunction of protection equipment or lightning. In a system without redundancy a fault often leads to a long interruption, which requires manual intervention. Short interruptions are often caused by automatic reclosing after a fault. Short interruptions below three minutes are normally considered a voltage quality problem. Interruptions are a severe power quality problem, but in a wide range of industrial countries interruptions occur very rare, because of redundancy and high maintenance of the grid (Nielsen et al., 2002)

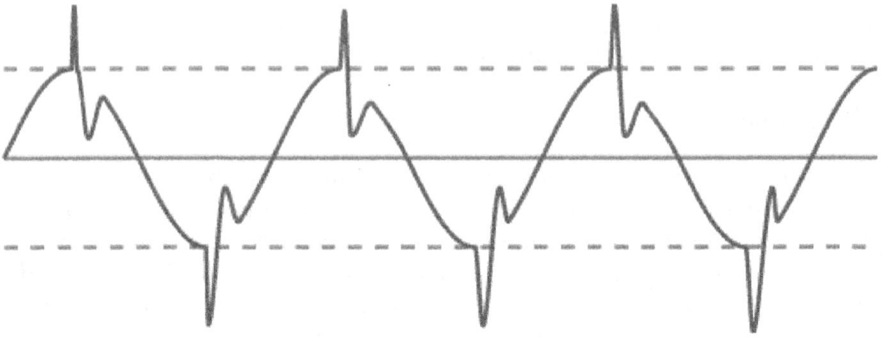

Figure 2.9: Interruption

*****

CHAPTER 3
POWER QUALITY SOLUTIONS

3. Introduction

Most of the more important international standards define power quality as the electric supply protected from any deviation in voltage, current or frequency under normal conditions & don't disrupt or disturb the customer's processes. Quality of power supply is basically defined by its two pivotal factors, voltage quality & supply reliability. These pivotal factors lead to power quality problems when they suffer change in their characteristics due to equipment failure or sudden system disturbances. The increasing trend of non-linear loads in the power system network has lead to the hazardous consequences suffered by the customer. This has further resulted in the slew of forward looking enhancements in system protection using Custom Power devices. They aptly enhance the stability & reliability of the system by reducing the effects of various disturbances

3.1 Definition of Power Quality

The quality assurance of electric power demands a deep research and study on the subject "Electric Power Quality". The interesting discussions & their respective proposals lead to various definitions of "Power Quality". It is term which different people described in different ways as follows:

1. Institute of Electrical and Electronic Engineers (IEEE) Standard

IEEE1100 defines power quality as "the concept of powering and grounding sensitive electronic equipment in a manner suitable for the equipment." 2. Electric Power Quality (EPQ) is a term that refers to maintaining the near sinusoidal waveform of power distribution bus voltages and currents at rated magnitude and frequency. Thus EPQ is often used to express voltage quality, current quality, reliability of service, quality of power supply, etc. 3. Power Quality is defined as the set of Parameters defining the properties of the power supply as delivered to the user in normal operating conditions in terms of continuity of supply and characteristics of voltage (symmetry, frequency, magnitude, waveform)

3.2 Power Quality Problems

The users demand higher power quality to use more sensitive loads to automate processes and improve living standards. Some basic criterions for power quality are constant rms value, constant frequency, symmetrical three-phases, pure sinusoidal wave shape and limited THD. These values should be kept between limits determined by standards if the power quality level is considered to be high. The economic losses due to power interruptions and disturbances can be quite high as a result of the important processes controlled and maintained by the sensitive devices. Power quality disturbances can be summarized as follows:

1. User loads
2. User electrical system and grounding
3. Weather related such as lightening, wind and rain
4. Utility distribution system
5. Utility transmission system
6. Utility generation system

1. **Interruption/under voltage/over voltage:** They are among the very common type of disturbances. During power interruption, voltage level of a particular bus goes down to zero. The interruption may occur for short or medium or long period. Such disturbances are increasing the amount of reactive power drawn or deliver by a system, insulation problems and voltage stability.
2. **Voltage/Current unbalance:** Voltage and current unbalance may occur due to the unbalance in drop in the generating system or transmission system and unbalanced loading. During unbalance, negative sequence components appear. It hampers system performance and in some cases it may hamper voltage stability.
3. **Power system harmonics:** They are low-frequency phenomena characterized by waveform distortion, which introduces harmonic

frequency components. Voltage and current harmonics have undesirable effects on power system operation and power system components. In some instances, interaction between the harmonics and the power system parameters (R–L–C) can cause harmonics to multiply with severe consequences.

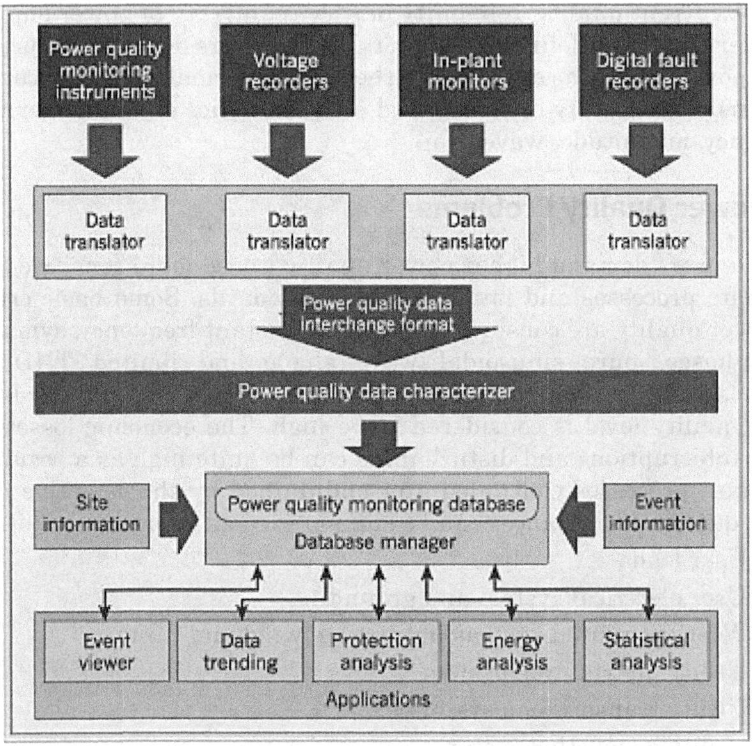

Figure 3.2: Power Quality Problems

4. **Power frequency disturbances:** These are low-frequency phenomena that result in voltage sags or swells. These may be source or load generated due to faults or switching operations in a power system.
5. **Power system transients:** They are fast, short-duration events that produce distortions such as notching, ringing, and impulse. The mechanisms by which transient energy is propagated in power lines, transferred to other electrical circuits, and eventually dissipated are different from the factors that affect power frequency disturbances.
6. **Voltage sag:** It is a short duration disturbance. During voltage sag, r. m. s. voltage falls to a very low level for short period of time.

7. **Voltage swell:** It is a short duration disturbance. During voltage sag, r. m. s. voltage increases to a very high level for short period of time.
8. **Flicker:** It is visual effect and undesirable frequency variation of voltage in a system.
9. **Ringing waves:** An oscillatory disturbance of decaying magnitude for short period of time is known as ringing wave. It may be called a special type transient.
10. **Outage:** It is special type of interruption where power cut has occurred for not more than 60 s due to fault or mal-tripping of switchgear/system

3.3 Quality Solutions

The integrated solution to the power quality problems encountered in the distribution area is "Custom Power". It focuses on reliability &quality of power flow. Custom power technology, the low-voltage counterpart of the more widely known flexible ac transmission system (FACTS) technology, aimed at high-voltage power transmission applications, has emerged as a credible solution to solve many of the power quality problems relating to continuity of supply at the end-user level [4]. Mitigation is in the harmonic context which is often seen as reduction of harmonic voltage or current distortion.

- Reducing the power supply disturbances
- Improving the load equipment immunity to disturbances
- Inserting corrective equipment between the electrical supply and the sensitive loads

Figure 3.3: Power Quality Solutions

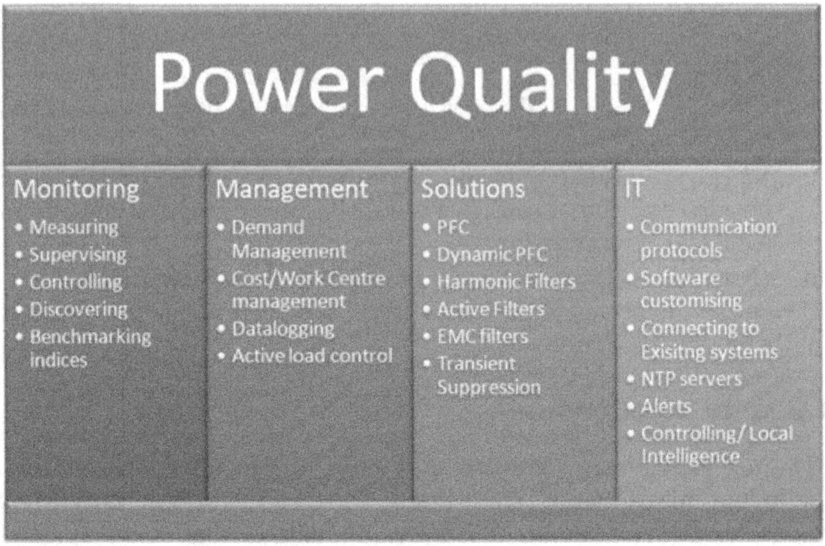

Figure 3.3.1: Power quality

3.4 Power Quality Standards

Math Bollen et al (2000) discuss about power quality improvement methods and solutions, which can be implemented independently of any standardization. However, proper standardization provides important guidelines for the implementation of the technical solutions. The power quality standards are mostly concerned with the following three areas: Defining the nominal environment Defining the terminology Limiting the number of power quality problems The International Electro technical Commission (IEC) and the Institute of Electrical and Electronics Engineers (IEEE) have proposed their sets of power quality standard. The highlights are given below: EN 50160 Voltage Characteristics of Public Distribution Systems Standard 519-1992, IEEE Recommended Practices and Requirements for Harmonic Control in Electrical Power Systems. A critical review of various standards is given in Math Bollen et al (2000), including the detailed description of the European voltage characteristic standard EN 50160. Proper standardization has to solve the problem of power quality disturbances adequately. Arindam Ghosh et al (2002) have listed various power quality standards as shown in Table 2.1.

*****

CHAPTER 4
DYNAMIC VOLTAGE RESTORER

4. Introduction

Dynamic Voltage Restoration (DVR) is a method and apparatus used to sustain, or restore, an operational electric load during sags, or spikes, in voltage supply. Often used in manufacturing areas requiring significant power to run tools/equipment, and utility plants, this custom device mitigates potential damage to equipment and undesirable slowdowns to the production line caused by an abrupt change in electric load. This method uses critical devices such as an automatic Transfer switch and IGBT Modules in order to operate. Dynamic Voltage Restorer and Its application at LV & MV Level DVR (Dynamic Voltage Restorer) is a static var device that has seen applications in a variety of transmission and distribution systems. It is a series compensation device, which protects sensitive electric load from power quality problems such as voltage sags, swells, unbalance and distortion through power electronic controllers that use voltage source converters (VSC). The first DVR was installed in North America in 1996 - a 12.47 kV system located in Anderson, South Carolina. Since then, DVRs have been applied to protect critical loads in utilities, semiconductor and food processing. Today, the dynamic voltage restorer is one of the most effective PQ devices in solving voltage sag problems. However, cost and installation restrictions have limited its implementation to where there is obvious requirement for a stable voltage supply. The basic principle of the dynamic voltage restorer is to inject a voltage of required magnitude and frequency, so that it can restore the load side voltage to the desired amplitude and waveform even when the source voltage is unbalanced or distorted. Generally, it employs a gate turn off thyristor (GTO) solid state power electronic switches in a pulse width modulated (PWM) inverter structure. The DVR can generate or absorb independently controllable real and reactive power at the load side. In other words, the DVR is made of a solid state DC to AC switching power converter that injects a set of three phase AC output voltages in series and synchronism with the distribution and transmission line voltages. The source of the injected voltage is the commutation process for reactive power demand and an energy source for the real power demand. The energy source may vary according to the design and manufacturer of the DVR. Some examples of energy sources applied are DC capacitors, batteries and that drawn from the line through a rectifier. Power quality is an issue; gaining significant interest due to huge economic losses which is the result of various voltage disturbances occurring in the system. Among the voltage disturbances, voltage sag is most severe that adversely affects the performance of the system. The one such efficient & reliable solution is the DVR. DVR is a static series compensator that injects voltage in series to the distribution system, regulating the load side voltage. It is connected between the supply and the sensitive load to compensate the line voltage harmonics, reduction of transients in addition to compensation of voltage sags & swells.

A power electronic converter based series compensator that can protect critical loads from all supply side disturbances other than outages is called a dynamic voltage restorer. The restorer is capable of generating or absorbing independently controllable real and reactive power at its AC output terminal. This device employs solid-state power electronic switches in a pulse-width modulated (PWM) inverter structure. It injects a set of three-phase AC output voltages in series and synchronism with the distribution feeder voltages. The amplitude and phase angle of the injected voltages are variable thereby allowing control of the real and reactive power exchange between the device and the distribution system. The DC input terminal of the restorer is connected to an energy source or an energy storage device of appropriate capacity. The reactive power exchanged between the restorer and the distribution system is internally generated by the restorer without AC passive reactive components. The real power exchanged at the restorer output AC terminals is provided by the restorer input DC terminal from an external energy source or energy storage system [7].]

The series controllers for control of the fundamental voltage are termed as a series connected PWM regulator in, a static series regulator in and, but mostly the devices are termed dynamic voltage restorers. If the device only injects reactive power the device can be termed as series var compensators. Taking the same simplified model of supply and load, but now with a series controller inserted to support the load. A 0.5 pu voltage sag can by a series device be restored by a 0.5 pu DVR and only 0.5pu of the energy absorbed by the load has to be supplied by the DVR. The supply continues to be connected and no resynchronization is necessary as it is the case with a shunt connected converter. The series voltage controller is connected in series with the protected load as shown in figure 3.3. Usually the connection is made via a transformer, but configurations with direct connection via power electronics also exist. The resulting voltage at the load bus bar equals the sum of the grid voltage and the injected voltage from the DVR. The converter generates the reactive power needed while the active power is taken from the energy storage. The energy storage can be different depending on the needs of compensating. The DVR often has limitations on the depth and duration of the voltage sag that it can compensate. Therefore right sized has to be used in order to achieve the desired protection. Options available for energy storage during voltage sags are conventional capacitors for very short durations but deep, batteries for longer but less severe magnitude drops and super capacitors in between. There are also other combinations and configurations possible. There are configurations, which can work without any energy storage, and they inject a lagging voltage with the load current. There are also different approaches on what to inject to obtain the most powerful solution. The main advantage with this method is that a single DVR can be installed to protect the whole plant (a few MVA) as well as single loads. Because of the fast switches, usually IGBT's, voltage compensation can be achieved in less than half a cycle. Disadvantages are that it is relatively expensive and it only mitigates voltage sags from outside the site. The cost of a DVR mainly depends on the power rating and the energy storage capacity

4.1 Principle of DVR

The main aim of DVR is to regulate the voltage at the load terminals irrespective of sag, distortion or unbalance in the supply voltage. The basic operating principle is to inject a voltage of required magnitude & frequency to restore the load voltage under voltage sag or distortion. Generally; it employs solid state power electronic switches such as GTO, IGBT or IGCT in the VSI, which can be operated in various pulse width modulation techniques such as SPWM(sinusoidal pulse width modulation),MSPWM(multiple sinusoidal pulse width modulation) .They inject a set of three phase AC voltage in series & synchronism with the distribution system. In normal condition it operates in the standby mode. During the disturbance, the nominal voltage is compared with the voltage variation in order to calculate the voltage to be injected by the DVR to maintain the supply voltage within limits. The DVR is capable of providing the reactive power compensation but the real power is provided by the energy storage system. The DVR functions by injecting three single phase AC voltages in series with the three phase incoming network voltages during a sag, compensating for the difference between faulty and nominal voltages. All three phases of the injected voltages are of 11 controllable amplitude and phase. Three pulse-width modulated (PWM) voltage source inverters (VSI) fed from a DC link supply the active and reactive power. During undisturbed power supply condition, the DVR operates in a low loss standby mode. In the normal operation mode (no sag) the low voltage side of the booster transformer is shorted either by solid state bypass switch or by switching one of the inverter legs and it functions as a short-circuited current transformer. Since no VSI switching takes place, the DVR produces conduction losses only. These losses should be kept as low as possible so as not to cause steady state power loss. Harmonics produced by the operation of VSI must be reduced to an acceptable limit defined by [11] by proper filtering scheme. Modulation scheme used on the VSI switches has also impact on the harmonics produced. The required energy during sags has to be supplied by an energy source. The necessary amount of energy that must be delivered by the energy source depends on load MVA requirement, control strategy applied, deepest sag to be protected. Under normal conditions, the short circuit impedance of the injection transformer determines the voltage drop across the DVR. This impedance must be low and has an impact on the fault current through the VSI on secondary side caused by a short circuit at load side. The filter design is also affected by the impedance of the injection transformer. In case of fault or over current exceeding the rating of DVR on the load side, solid state bypass switches or electromechanical bypass switches must be added as a measure to protect DVR from getting damaged. As an overview the following main design criteria influence the rating and the performance of the DVR [5]

4.2 Operation

In normal conditions, the dynamic voltage restorer operates in stand-by mode. However, during disturbances, nominal system voltage will be compared to the voltage variation. This is to get the differential voltage that should be injected by the DVR in order to maintain supply voltage to the load within limits. The amplitude and phase angle of the injected voltages are variable, thereby allowing control of the real and reactive power exchange between the dynamic voltage restorer and the distribution system. The DC input terminal of a DVR is connected to an energy storage device of appropriate capacity. As mentioned, the reactive power exchange between the DVR and the distribution system is internally generated by the DVR without AC passive reactive components. The real power exchanged at the DVR output AC terminals is provided by the DVR input DC terminal by an external energy source or energy storage system. Also, there is a resemblance in the technical approach to DVRs to that of providing low voltage ride-through (LVRT) capability in wind turbine generators. The dynamic response characteristics, particularly for line supplied DVRs are similar to LVRT-mitigated turbines. Moreover, since the device is connected in series, there are conduction losses, which can be minimized by using Integrated Gate-Commutated Thyristor (IGCT) technology in the inverters

Maximum MVA-load and power factor, 12

1. Maximum 1 phase and 3 phase voltage sags to be compensated,
2. Maximum duration of 3 phase voltage sag,

3. Maximum allowed voltage drop across DVR under steady-state conditions,
4. Short circuit impedance of the injection transformers,
5. Short circuit impedance and connection of step down transformers at the input and output sides of the DVR as well as the short circuit power
6. DVR Components A typical DVR consist of the following major components [5]:
7. Voltage Source Inverter/s (VSI)
8. Injection Transformer/s
9. Harmonic Filter
10. Energy Storage Unit
11. DC Link
12. Control and Protection System

4.3 Operation of DVR

The DVR can be operated in three different modes which are described as:

1. Bypass mode: the Dynamic Voltage Restorer (DVR) is bypassed mechanically or electronically during high load currents and downstream short circuits. In this mode the DVR cannot inject a voltage to improve the voltage quality.

2. Standby mode: The supply voltages are at rated level and the DVR is ready to compensate for voltage sag. During standby mode the DVR can have secondary tasks and operation modes.

3. Loss less mode: The DVR performs no switching and the losses in the DVR are minimized to conduction losses. Harmonic blocking mode and voltage balancing mode: – The DVR improves the load voltage and compensate for poor background voltage quality. The DVR has to perform switching and is expected to inject a relatively small voltage.

3. Capacitor emulation mode: The DVR is controlled to; – operate as in inserted series capacitor, thereby it can compensate for large line inductance and for inductance inserter in conjunction with the DVR.

4. Active mode: Whenever voltage sags are detected, DVR injects the missing voltage. In this mode DVR should ensure the unchanged load voltage with minimum energy dissipation for injection due to high cost of capacitors. The available voltage injection strategies are pre-sag, phase advance, voltage tolerance and in phase method.

4.4 Caution

DVRs may provide good solutions for end-users subject to unwanted power quality disturbances. However, there is a caution regarding their application in systems that are subject to prolonged reactive power deficiencies (resulting in low voltage conditions) and in systems that are vulnerable to voltage collapse. In many cases, the main protection of the power system against voltage collapse is the natural response of load to decrease demand when system voltage drops. The application of DVRs would tend to maintain demand even when incipient voltage conditions are present. As a result, this reduces the innate ability to prevent a collapse and increases the chance of cascading interruptions. In addition, from the transmission viewpoint, the dynamic voltage restorer would extend the voltage range if the load is a constant power type. The combination of direct-connected DVRs, voltage-switched capacitor banks and on-load tap-changing distribution transformers, leads to more current drawn from the transmission system during periods of reactive deficiency and low voltages. Therefore, when applying DVRs, it is vital to consider the nature of the load whose voltage supply is being secured, as well as the transmission system which must tolerate the change in voltage-response of the load. It may be necessary to provide local fast reactive supply sources in order to protect the system, with the DVR added, from voltage collapse and cascading interruptions. A comprehensive simulation study, which includes the transmission system, is highly recommended.

4.5 Voltage Injection Methods

Since the dynamic voltage restorer injects the compensating voltage in order to maintain the load voltage constant, there are certain limitations in compensating the voltage sags. The factors influencing the compensation are finite power rating, different load conditions and different types of voltage sag. Load characteristics dictate the control strategy of dynamic voltage restorer as some loads are sensitive to phase angle jump and others are tolerant to phase angle jump. The injection compensating voltage is categorized as three methods.

1. Pre-sag Compensation: In this method DVR continuously tracks the supply voltage. The DVR injects the missing voltage between during sag and pre-sag voltages to the system. During the compensation, DVR has to compensate both magnitude and angle. The Fig 3.6 shows the vector representation of pre-sag conditions. In this method, the injected power cannot be controlled while load voltage can be restored ideally. Load conditions and type of fault determines the injected power.

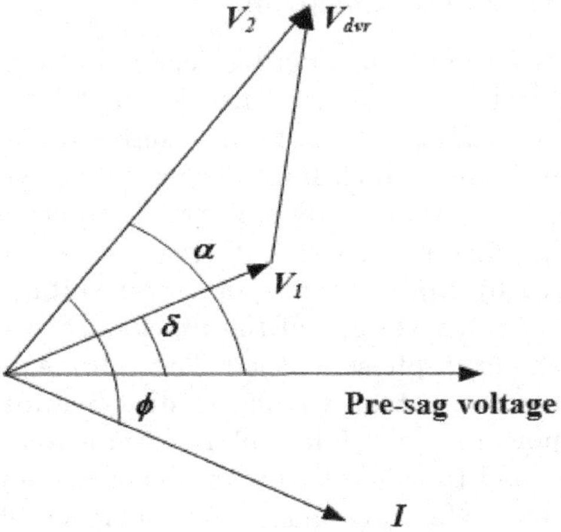

Figure 4.5: Voltage Injection Methods

4.5.1 In-phase Compensation

In this method, injected voltage is independent of the load current and pre-fault voltage. The injected voltage is always in-phase with supply voltage. The added advantage of this method is to minimize the magnitude of injected voltage for constant load voltage magnitude. The phase angles of pre-sag and load voltage are different and magnitude of load voltage is same as pre-fault voltage. The Fig 3.7 represents the in-phase compensation. During normal conditions, supply voltage and load voltage are equal with zero phase angle. During abnormal condition like sag, phase angle jump appears with reduction of load voltage magnitude. Both the parameters are compensated by DVR. Fig 3.7: Vector diagram

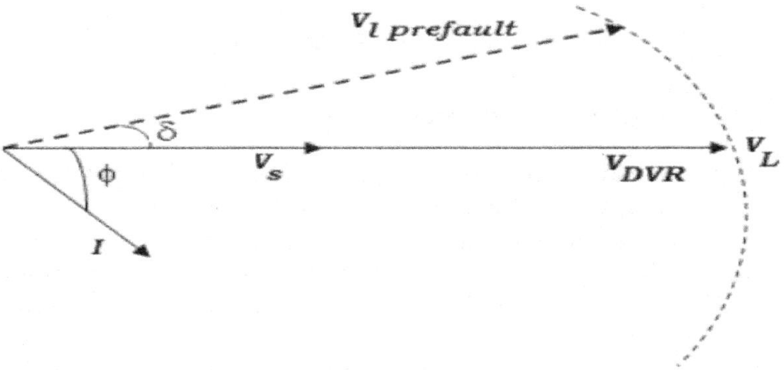

Figure 4.5.1: In-phase compensation

4.5.2 Phase-Advance Compensation

Active power is injected to sensitive loads continuously in both pre-sag compensation and In-phase compensation methods. In both methods, DVR restoration time and performance are confined due to limited energy storage capacity of dc link which limits the active power injection. Phase advance methods proves to better compared to other methods as it associates only reactive power injection instead of active power. The injected active power is made zero by injecting compensating voltage perpendicular to load current. The injected voltage of the phase-advance method is larger than those of pre-sag or in-phase method. The voltage phase shift in phase advance method causes voltage waveform discontinuity, inaccurate zero crossing and load power swing. Hence, phase advance compensation should be adjusted to the load that is tolerant to phase angle jump or transition period should be taken while the phase angle is moved from pre-fault angle to the advance angle.

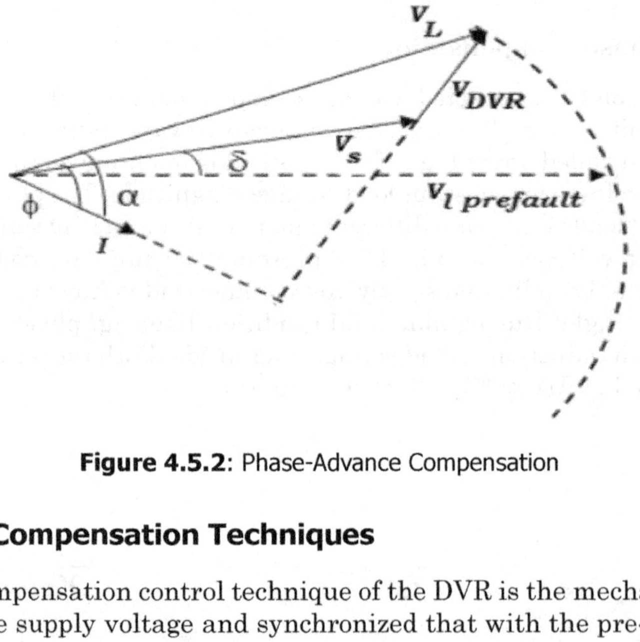

Figure 4.5.2: Phase-Advance Compensation

4.6 DVR Compensation Techniques

The compensation control technique of the DVR is the mechanism used to track the supply voltage and synchronized that with the pre-sag supply voltage during a voltage sag/swell in the upstream of distribution line. Generally voltage sags are associated with a phase angle jump in addition to the magnitude change [21]. Therefore the control technique adopted should be capable of compensating for voltage magnitude, phase shift and thus the wave shape. But depending on the sensitivity of the load connected downstream, the level of compensation of the above parameters can be altered. Basically the type of load connected influences the compensation strategy. For example, for a linear load, only magnitude compensation is required as linear loads are not sensitive to phase angle changes [11, 13]. Further when deciding a suitable control technique for a particular load it should be considered the limitations of the voltage injection capability (i.e. the rating of the inverter and the transformer) and the size of the energy storage device [11]. Compensation is achieved via real power and reactive power injection. Depending on the level of compensation required by the load, three types of compensation methods are defined and discussed below namely pre-sag compensation, in-phase compensation and energy optimization technique. The circuit for a simple power system with a DVR is shown in Figure 2.9 below. The supply voltage, Load voltage, Load current and the voltage injected by the When the system is in normal condition; the supply voltage (Vs) is identified as pre-sag voltage and denoted by Vpre-sag. In such situation since the DVR is not injecting any voltage to the system, load voltage (Vload) and the supply voltage will be the same. During voltage sag, the magnitude and the phase angle of the supply voltage can be changed and it is denoted by Vsag. The DVR is in operative in this case and the voltage injected will be VDVR. If the voltage sag is fully compensated by the DVR, the load voltage during the voltage sag will be Vpre-sag.

4.7 Control Techniques Used in Commercially Available DVRs

Most of the commercially available DVRs use either the in-phase compensation technique or energy optimization technique, owing to minimal requirement of real power injection: hence it reduces the capacity of the energy storage needed. Control technique describes the method used to quantify the DVR control voltage injected during the compensation. In simple terms it basically detects the occurrence of voltage sag. Some common control techniques used by DVR

Manufacturers are described in this section [11]. Irrespective of the compensation techniques used, there should be a scheme to track the phase angle and the magnitude of the supply voltage during normal operation (more specifically positive sequence component of the supply voltage) and to detect the occurrence of voltage sag. In other words there should be a voltage sag detection technique (it detects the occurrence of the sag, start and end points sag depth and phase shift). Followings are some of the common voltage sag detection techniques.

4.7.1 Voltage Sag Detection Techniques

1. Fourier transforms
2. Phase Locked Loop (PLL)

3. Vector control (Software Phase Locked Loop –SPLL)
4. Peak value detection
5. Applying the wavelet transform to each phase

Out of the techniques mentioned above only the Fourier transform, Vector control and wavelet transform methods provide both the voltage magnitude and phase shift information. PLL method can provide only the phase shift information while peak value detection technique enables to get the magnitude change (voltage sag) information. Hence it is possible to combine one or more techniques mentioned above to obtain accurate voltage sag compensation.

4.8 Location of DVR

The intention is only to protect one consumer or a group of consumers with value added power. Applying a DVR in the medium or low voltage distribution system would often be possible and a radial grid structure is the only type of system considered here. In Europe three wire systems are common in the medium voltage systems and four wires in low voltage systems. In both systems the main purpose is to inject synchronous voltages during symmetrical faults and in some cases inject an inverse voltage component during non-symmetrical faults [9]. A main difference between a Low Voltage (LV) connection and a Medium Voltage (MV) connection is the flow of zero sequence currents and the generation of zero 13 sequence voltages. In the four-wire system, the DVR must secure a low impedance for zero sequence currents and the zero sequence must either flow in the power converter or in a delta winding of the injection transformer.

4.9 Converter Type

Pulse width modulated voltage source inverters are commonly used converters for this type of application. Applying a VSI to compensate for a missing voltage requires an injection transformer and a line filter to damp generated switching harmonics. Additionally, losses should be minimized, because the DVR conducts the load current continuously and a fast response is important to avoid a deterioration of the load voltage. The critical factors when choosing a converter for a DVR application are [9]: Instructional

- The circuit for single-phase Current Source Inverter (CSI) using thyristors
- Auto-Sequential Commutated mode of operation for 1-ph. Inverter (ASCI), with waveforms
- Three-phase Current Source Inverter (CSI) – circuit and operation, with waveforms

4.9 Introduction

In the previous six (5.1-5.6) lessons in this module, the circuit and operation of single-phase and three-phase Voltage Source Inverters (VSI), with waveforms, were described in detail. Also, the presence of harmonics in voltage waveforms, along with its reduction mainly by Pulse Width Modulation (PWM) techniques, was presented. Presently, mainly self-commutated switching devices, like say transistors, are used in the above circuits, replacing thyristors, with bulky commutation circuits needed to turn them OFF, these being force-commutated ones. In the last two (5.7-5.8) lessons in this module, the circuit and operation of different types of single-phase and three-phase Current Source Inverters (CSI), with waveforms, will be described in detail. The device used here is thyristor. In this lesson (5.7), initially, the circuit of single-phase CSI will be presented. The Auto-Sequential Commutated mode of operation for this Inverter (ASCI), using thyristors, will be discussed in detail, with waveforms. Then, the circuit and operation of threephase CSI, along with relevant waveforms, will be presented. Finally, the advantages and disadvantages of CSI over VSI, in brief, are described For the VSI, as the full form denotes, the output voltage is constant, with the output current changing with the load - type, and/or the values of the components. But in the CSI, the current is nearly constant. The voltage changes here, as the load is changed. In an Induction motor, the developed torque changes with the change in the load torque, the speed being constant, with no acceleration/deceleration. The input current in the motor also changes, with the input voltage being constant. So, the CSI, where current, but not the voltage, is the main point of interest, is used to drive such motors, with the load torque changing.

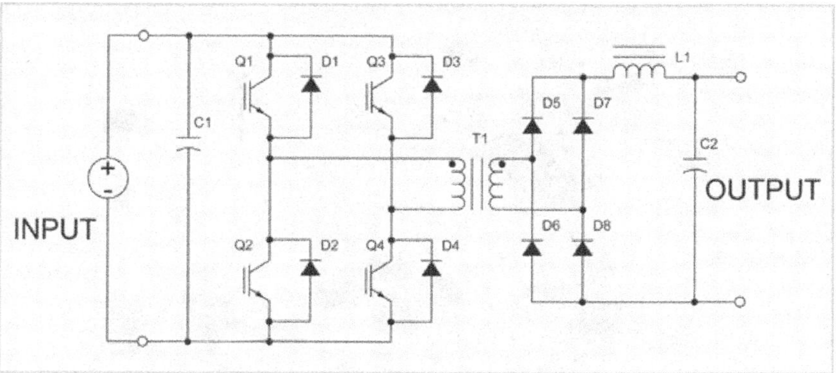

Figure 4.9: Voltage Source Inverters

1. Switching frequency
   - Generation of switching harmonics

- Ripple current in the converter
- Power losses caused by the switching and conduction

2. Dynamic response
3. DVR impedance

4.10 Energy Source

Compensating effective for large voltage reductions requires active power, either from a storage system, the grid itself or from an auxiliary supply. The solutions can be summarized as follows [9]: 14

1. Energy storage
2. Auxiliary supply
3. Line connected shunt converter
   - Load connected
   - Source connected

Operating with energy storage gives limitations in the performance to compensate for long duration sags and large energy storage is still costly, therefore most of the proposed DVR solutions have been realized with a shunt converter fed from the line itself or an auxiliary supply. Main advantage of load side connected shunt converter is that the shunt converter has a restored clean voltage to enable a stable charging to the common DC link. A drawback with the load side connected is the requirement of a higher current rating of the series converter [9].

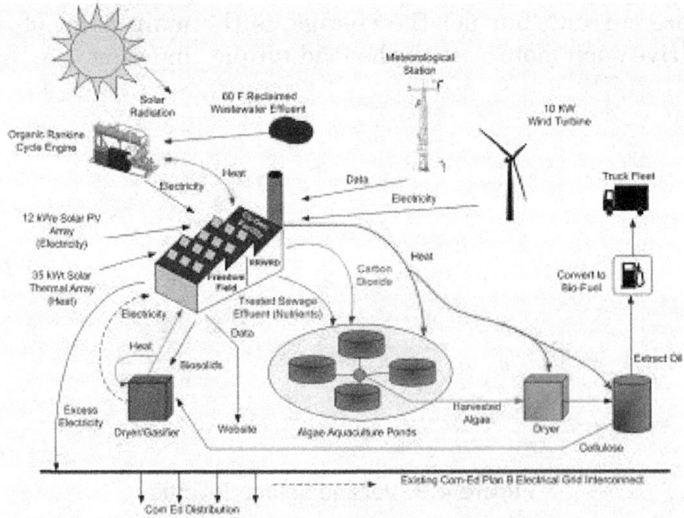

Figure 4.10: Energy storage

4.11 Injection Transformer

The injected voltages are introduced into the distribution system through an injection transformer connected in series with the distribution feeder. It is known that in order to guarantee the maximum reliability and effectiveness of this restoration scheme, clever selection of the injected transformer is a prerequisite. In order to carefully select a suitable injection transformer the following issues should carefully be addressed [17]: 15 Transformers are commonly used in applications which require the conversion of AC voltage from one voltage level to another. There are two broad categories of transformers: electronic transformers, which operate at very low power levels, and power transformers, which process thousands of watts of power. Electronic transformers are used in consumer electronic equipment like television sets, VCRs, CD players, personal computers, and many other devices, to reduce the level of voltage from 220V (available from the AC mains) to the desired level at which the device operates. Power transformers are used in power generation, transmission and distribution systems to raise or lower the level of voltage to the desired levels. The basic principle of operation of both types of transformers is the same. In this chapter, we will first review some of the basic concepts of magnetic circuits, which are fundamental building blocks in transformers and electric machinery. In order to understand how a transformer operates, we will examine two inductors that are placed in close proximity to one another. The concepts of such magnetic coupled circuits will be extended to the development of transformers. After understating the relationships between voltages and currents, we will look at some practical considerations regarding the use of transformers.

4.11 Injection Transformer

- The MVA rating
- The primary winding voltage and current ratings
- The turn-ratio which, in turn, determines the secondary winding voltage and current rating
- The short-circuit impedance

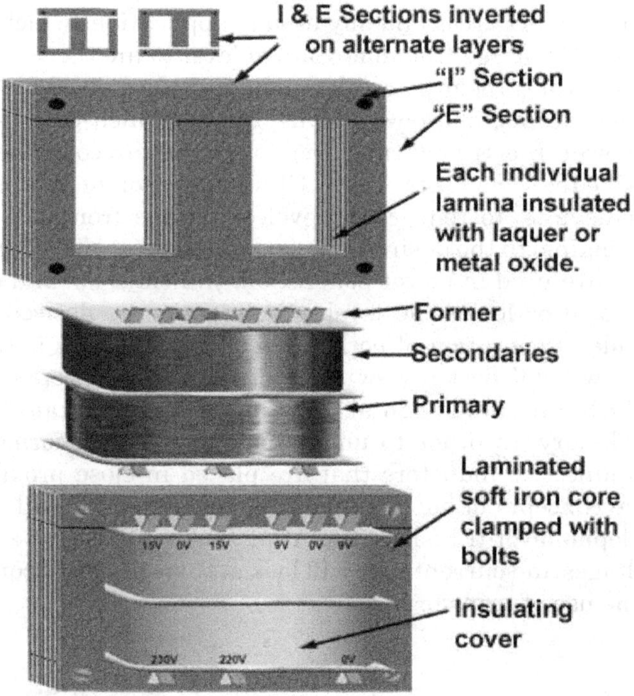

Figure 4.11.1: Winding

4.12 Filtering Schemes

In the design of DVR, special attentions must be paid to the filtering scheme as it is related with the system dynamic response. Two filtering schemes were seen from the recent literature, which are the inverter-side filter and line-side filter [12], [13]. Figure 2.4 shows the configuration of the inverter-side filter, and Figure 2.5 depicts that of the line-side filter. In the scheme of inverter-side filter, a series inductor Lf and a shunt capacitor Cf are inserted between the transformer and switching devices.

From the view point of harmonic elimination, the inverter-side filter scheme may be a preferable one. Yet, as the insertion of the inductor may introduce an additional voltage-drop component across the series transformer, it is also found that the magnitude and phase difference between the fundamental components of voltage has appeared on both sides of the filter, implying that an inadvertent choice of the filter design would significantly downgrade the DVR control performance. As for the DVR design of line-side filter, it can be implemented by placing the filtering scheme on the line-side of the series transformer as Figure 2.5 depicts. In such a line-side filter scheme, the leakage flux of series transformer is served as the filtering inductance, and the capacitor is placed across the transformer winding. By use of this design, the system will no longer require a physical inductor, avoiding the problem encountered in the inverter-side filter. However, similar to the inverter-side filter scheme, a transport delay is also encountered between the input and output of the filter network that may be attributed to the utilization of the line-side filter. Note that while the DVR along with the filtering scheme is devoted to cancel out the incoming nuisance, this resultant transport delay may instead offset the dynamic response in a significant manner, or even lead to a system compensation failure [14].

*****

CHAPTER - 5
VOLTAGE SOURCE INVERTER

5. VOLTAGE SOURCE INVERTER

A voltage source converter (VSC) is a power electronic device, which can generate a three-phase ac output voltage is controllable in phase and magnitude [1]. These voltages are injected into the ac distribution system in order to maintain the load voltage at the desired voltage reference. VSCs are widely used in adjustable speed drives, but can also be used to mitigate the voltage sags and swells. The VSC is used to either completely replacing the voltage or to inject the 'missing voltage'. The 'missing voltage' is the difference between the nominal voltage and the actual voltage. The converter is normally based on the some kind of energy storage, which will supply the converter with a dc voltage.

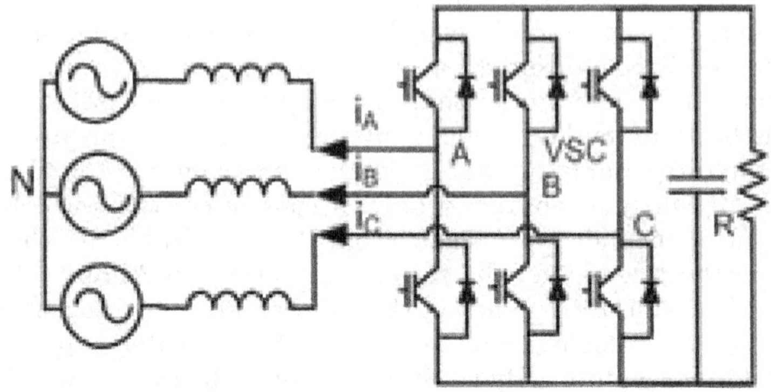

Figure 5.1: Voltage source converter (VSC)

Normally the VSC [6] is not only used for voltage dip mitigation, but also for other power quality issues, e.g. flicker and harmonics. A special gate unit and voltage divider across each IGBT maintain an even voltage distribution across the series connected IGBTs. The gate unit not only maintains proper voltage sharing within the valve during normal switching conditions but also during system disturbances and fault conditions. A reliable short circuit failure mode exists for individual IGBTs within each valve position.[9] Depending on the converter rating, series-connected IGBT valves are arranged in either a three-phase two-level or three-level bridge.

In three-level converters, IGBT valves may also be used in place of diodes for neutral point clamping. Each IGBT position is individually controlled and monitored via fiber optics and equipped with integrated ant parallel, free-wheeling diodes. Each IGBT has a rated voltage of It forms the building block of compensating device. It performs the power conversion process from DC to AC. VSI consists of fully controlled semiconductor power switches to form a single phase or three phase topologies. For medium power inverters, IGBT's are used and GTO's or IGCT's due to compact size & fast response for high power inverters are employed. The single phase VSI topology encompasses a low-range power applications and medium to high power applications are covered by the three phase topology [10]. Single phase VSI consists of four semiconductor switches (in 2 legs) to generate the ac output waveform as in fig.3.1.Three phase VSI is a six step bridge inverter that uses a minimum of six thyristors as in fig 3.2, where a step means a change in the firing from one thyristor to the next thyristor in proper sequence. For one cycle of 360 degree, each step is of 60 degree for a six step inverter.

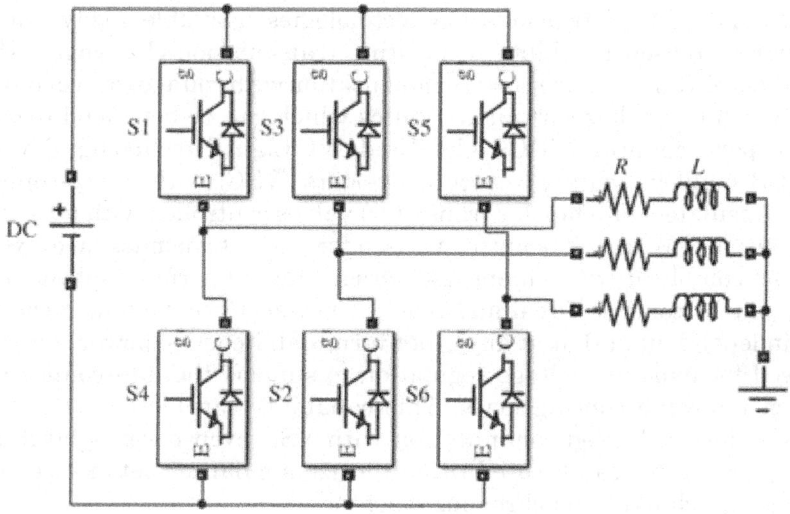

Figure 5.1.1: Six step inverter

2.5 kV with rated currents up to 1500 A Each VSC station is built up with modular valve housings which are constructed to shield electromagnetic interference (EMI). The valves are cooled with circulating water and water to air heat exchangers. PWM switching frequencies for the VSC typically range between 1-2 kHz depending on the converter topology, system frequency and specific application.

Voltage Source Converter (VSC) technology has been selected as the basis for several recent projects due to its controllability, compact modular design, ease of system interface and low environmental impact. This paper describes the rationale for selection of VSC technology and the latest technical developments utilized in several recent projects. Traditional HVDC and FACTS installations have often provided economic solutions for special transmission applications.HVDC is well-suited for long-distance, bulk power transmission, long submarine cable crossings, and asynchronous interconnections. Static var compensators (SVC) provide a reserve source of dynamic reactive power thereby raising power transfer limits. HVDC and FACTS technologies permit transmitting more power over fewer transmission lines. Deregulated generation markets, open access to transmission, formation of RTO's, regional differences in generation costs and increased difficulty in siting new transmission lines, however, have led to a renewed interest in FACTS and HVDC transmission often in non-traditional applications.

This is especially true with the lag in transmission investment and the separation in ownership of generation and transmission assets.

HVDC and FACTS transmission technologies available today offer the planner increased flexibility in meeting transmission challenges. HVDC transmission and reactive power compensation with voltage source converter (VSC) technology has certain attributes which can be beneficial to overall system performance. HVDC Light™ and SVC Light™ technology developed by ABB employs voltage source converters (VSC) with series-connected IGBT (insulated gate bipolar transistor) valves controlled with pulse width modulation (PWM) VSC converters used for power transmission (or voltage support combined with an energy storage source) permit continuous and independent control of real and reactive power. Reactive power control is also independent of that at any other terminal. Reactive power control can be used for dynamic voltage regulation to support the interconnecting ac system following contingencies. This capability can increase the overall transfer levels. Forced commutation with VSC even permits black start, i.e., the converter can be used to synthesize a balanced set of three phase voltages much like a synchronous machine.

A voltage-source converter is a power electronic device, which can generate a sinusoidal voltage with any required magnitude, frequency and phase angle. Voltage source converters are widely used in adjustable-speed drives, but can also be used to mitigate voltage dips. The VSC is used to either completely replace the voltage or to inject the 'missing voltage'. The 'missing voltage' is the difference between the nominal voltage and the actual. The converter is normally based on some kind of energy storage, which will supply the converter with a DC voltage. The solid-state electronics in the converter is then switched to get the desired output voltage. Normally the VSC is not only used for voltage dip mitigation, but also for other power quality issues, e.g. flicker and harmonics A special gate unit and voltage divider across each IGBT maintain an even voltage distribution across the series connected IGBTs. The gate unit not only maintains proper voltage sharing within the valve during normal switching conditions but also during system disturbances and fault conditions. A reliable short circuit failure mode exists for individual IGBTs within each valve position. Depending on the converter rating, series-connected IGBT valves are arranged in either a three-phase two-level or three-level bridge. In three-level converters, IGBT valves may also be used in place of diodes for neutral point clamping. Each IGBT position is individually controlled and monitored via fiber optics and equipped with integrated ant parallel, free-wheeling diodes. Each IGBT has a rated voltage of 2.5 kV with rated currents up to 1500 A. Each VSC station is built up with modular valve housings which are constructed to shield electromagnetic interference (EMI). The valves are cooled with circulating water and water to air heat exchangers. PWM switching frequencies for the VSC typically range between 1-2 kHz depending on the converter topology, system frequency and specific application

5.2 The Control of DVR

The control of a DVR is not straight forward because of the requirements of fast response, large variation in the type of sags to be compensated and variation in the type of connected load. The DVR must also be able to distinguish between background power problems and the voltage sags to be compensated. Sags are often nonsymmetrical and accompanied by a phase jump [15]. The possibility of compensation of voltage sags can be limited by a number of factors including finite DVR power rating, different load conditions, background power quality problems and different type of sags. If the DVR should be a successful 24 device, the control may be able to handle most sags and the performance must be maximized according to the equipment inserted. Otherwise, the DVR may not be able to avoid load tripping and even cause additional disturbance to load. A control strategy for voltage sags with phase jump should be included, to be able to compensate this particular type of sag. The control strategy can depend on the type of the load connected. Some loads are very sensitive to phase jump and the load should be protected from them. Other types of loads are more tolerant to phase jump and the main task is to maintain the nominal voltage on all three phases. Three basic control strategies for a DVR can be stated as [15]:

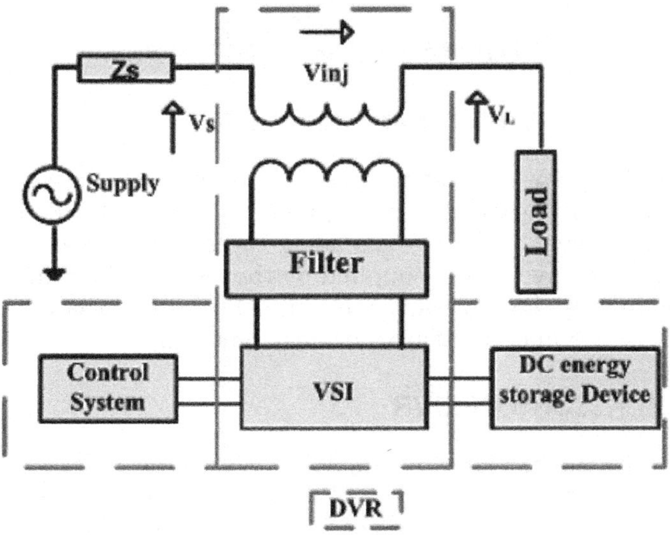

Figure 5.2: Basic control strategies for a DVR

5.3 PRE-SAG COMPENSATION METHOD

In this method, both magnitude & phase angle are to be compensated. The supply voltage is continuously tracked & load voltage is compensated to the pre-sag condition by injecting voltage equal to the difference of voltage under pre-sag & sag condition as in fig 3.3. Though, it gives a nearly undisturbed load voltage but suffers a drawback of exhausting the rating of the DVR.

5.4 IN PHASE COMPENSATION METHOD

In this method, when the source voltage drops due to sagging condition, the VSI injects a voltage called missing voltage based on the drop of voltage magnitude as in fig 3.5. The generated Voltage of the DVR is always in phase with the measured supply voltage regardless of the load current and the pre-sag voltage

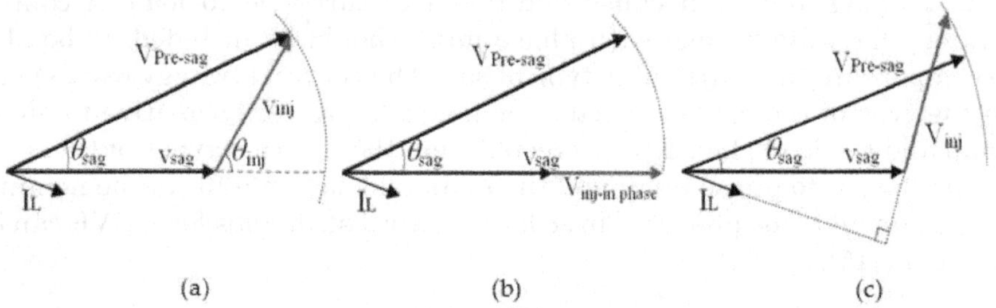

Figure 5.3: Compensation Method

5.5 REACTIVE POWER COMPENSATION

This is also known as the minimum energy injection, which depends on maximizing the active power supplied by the network (keeping the apparent power constant and decreasing the network reactive power) by minimizing the active power supplied by the compensator (increasing the reactive power supplied by the compensator). In this injection method the injected voltage is in quadrature with load current.

5.6 APPLICATIONS OF DVR

Practically, the capability of injection voltage by DVR system is 50% of nominal voltage. This allows DVRs to successfully provide protection against sags to 50% for durations of up to 0.1 seconds. Furthermore, most voltage sags rarely reach less than 50%. The dynamic voltage restorer is also used to mitigate the damaging effects of voltage swells, voltage unbalance and other waveform distortions. SSSC and DVR The SSSC's counterpart is the Dynamic Voltage Regulator (DVR). Although both are utilized for series voltage sag compensation, their operating principles differ from each other. The static synchronous series compensator injects a balance voltage in series with the transmission line. On the other hand, the DVR compensates the unbalance in supply voltage of different phases. Also, DVRs are usually installed on a critical feeder supplying the active power through DC energy storage and the required reactive power is generated internally without any means of DC storage.

The first DVR was installed in North America in 1996 - a 12.47 kV system located in Anderson, South Carolina [13]. Practically, the capability of injection voltage by DVR system is 50% of nominal voltage. This allows DVRs to successfully provide protection against sags to 50% for durations of up to 0.1 seconds. Furthermore, most voltage sags rarely reach less than 50%. The dynamic voltage restorer is also used to mitigate the damaging effects of voltage swells, voltage unbalance and other waveform distortions. DVRs of capacities up to 50 MVA have seen applications to critical loads in food processing, semiconductor and utility supply. Cost and installation constraints limit these to where there is clear need for constant voltage supply.

*****

CHAPTER 6
CONTROL PHILOSOPHIES OF DVR

6. Introduction

Voltage sags are one of the most severe power quality problems & DVR is an effective solution to mitigate it. The purpose of control scheme is to control the system output by generating an appropriate control signal prior to the unbalanced condition prevailing in the system. It generates the signals to enable the VSI (voltage source inverter) by providing proper firing sequence to the circuit. In this work, different control strategies for dynamic voltage restorer are investigated with emphasis on voltage sag compensation. Three promising control methods to compensate voltage sags are tested & compared with simulation of DVR on 11kV system. The comparison of the performance of three control strategies is made on basis of voltage waveforms & its frequency spectrum analysis. Their performance level is presented in the decreasing order of their compensation capability & better performance in mitigating voltage sag over a broader range for static linear & static non-linear load. Three control philosophies have been used namely, PI, Fuzzy & PI-Fuzzy. These are discussed as below: 4.2

6.1 Proprotional-Integral (Pi) Controller Based DVR PI

Is a feedback controller that uses the weighted sum of error & its integral value to perform the control operation? The proportional response can be adjusted by multiplying the error by constant Kp, called proportional gain. The contribution from integral term is proportional to both the magnitude of error and duration of error. The error is first multiplied by the integral gain, Ki and then was integrated to give an accumulated offset that have been corrected previously [8]. The input to the PI controller is difference between the reference value & error value of voltage. As per the comparison of reference value & error value of voltage, linear PI adjusts its proportional & integral gains Kp & Ki in order to reduce the steady state error to zero for a step input as shown in fig.4.1. It is widely used due to simple control structure but suffers a disadvantage of fixed gains i.e. it cannot adapt itself to the varying parameters & conditions of the system.

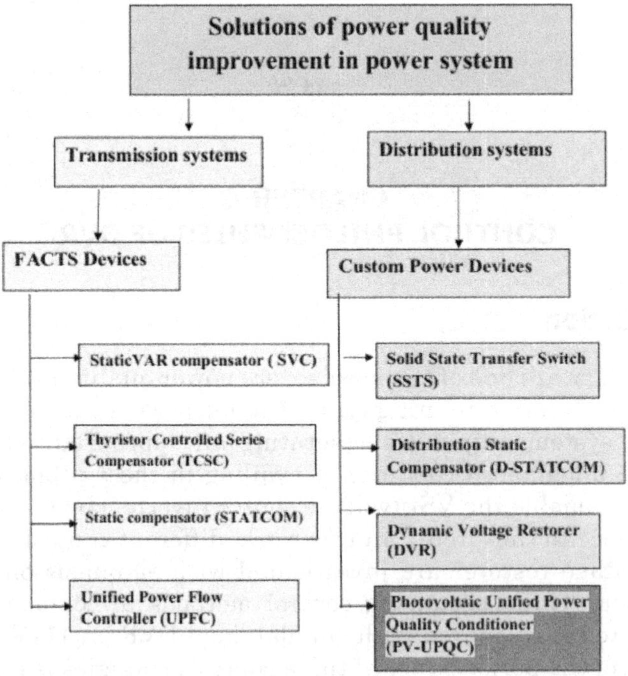

Fig. 4.1: Control Dvr

Advantage of Using DVR

Dynamic Voltage Restorer is the most efficient and effective modern custom power device used in power distribution network. Besides that, DVR also is lower cost, small size and its fast dynamic response to the disturbance.

Compensating Types

The compensating devices either compensate a load and correct its power factor or improve the quality of the supply voltage. The hierarchy of all the feasible power quality solutions along with the proposed are work highlighted as shown in Fig 4.1.

*****

CHAPTER 7
SIMULATION AND RESULT

7.1 Introduction

The main considerations for the control system of a DVR include: detection of the start and finish of the sag, voltage reference generation, transient and steady-state control of the injected voltage, and protection of the system. The control system was used to control the DVR with a sampling and switching frequency System performance is analyzed for compensating voltage sag with different DC storage capacity so as to achieve rated voltage at a given load. Various cases of different load condition are considered to study [4] the impact DC storage on sag compensation

System Quantities Standards

Source 3 phase, = 25e3 KV, 50 Hz

Inverter Parameters = IGBT based, 3 arms, 6 pulse, Carrier frequency= 1080 Hz, Sample time = 50 μs

PID Controller (2)= Kp1=20, Ki1=154, Sample time = 50 μs Kp2=25, Ki2=260, K_d=.001 Sample time = 50 μs

RL load Active power= 1KW, Inductive Reactive power= 500 VAR

Two winding transformer= Yg/Δ, 11/11KV

Transmission Line Parameter R= .001omh, L= 5e-3H

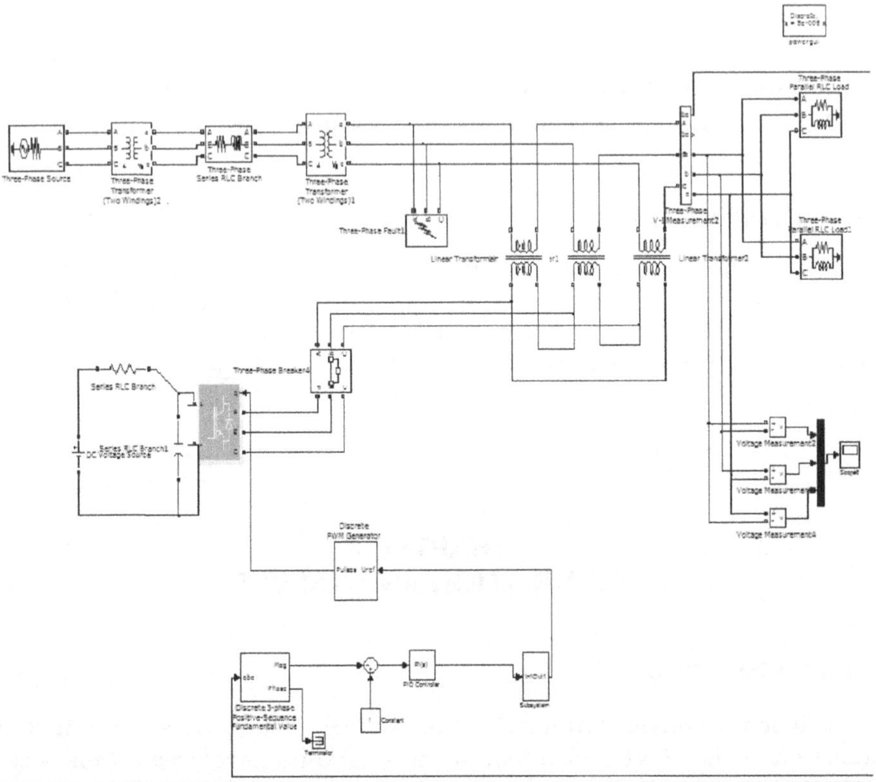

Figure 7.1: DVR Simulation

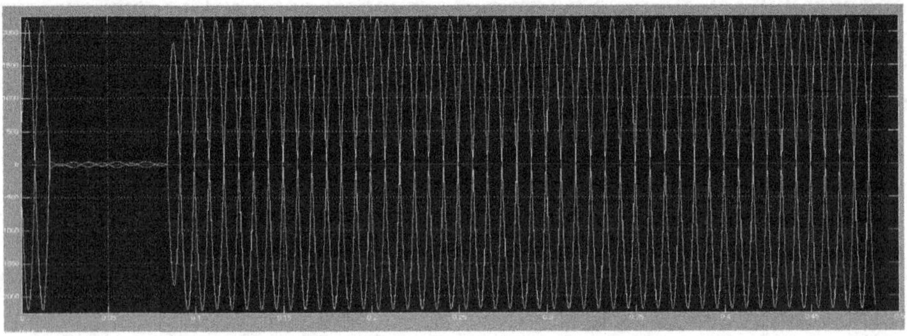

Figure 5: Three phase output voltage without DVR

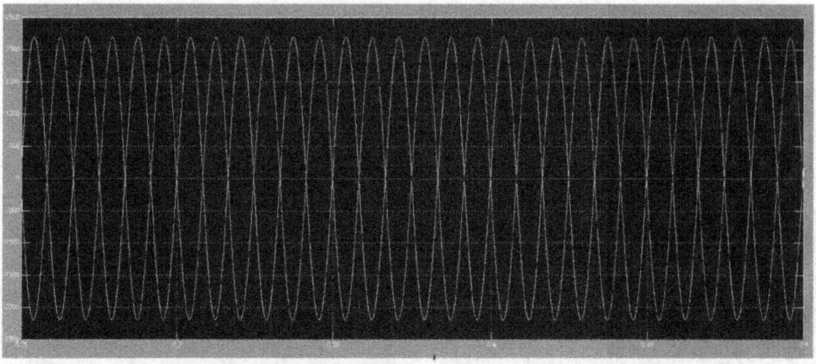

Figure 6: Three phase output voltage with DVR

Figure 7.2: PWM

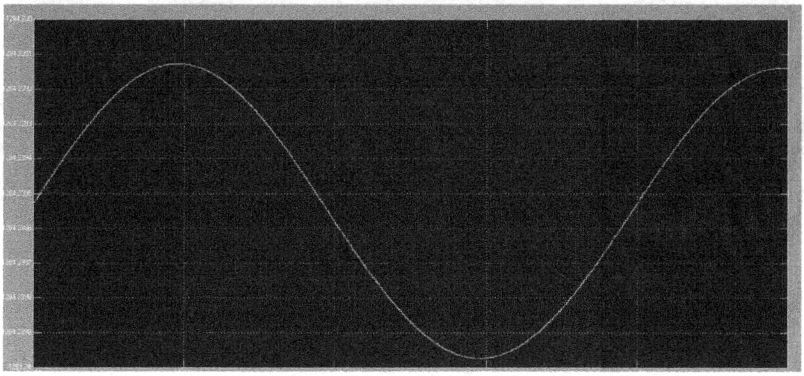

Figure 7.3: PI Controller Output

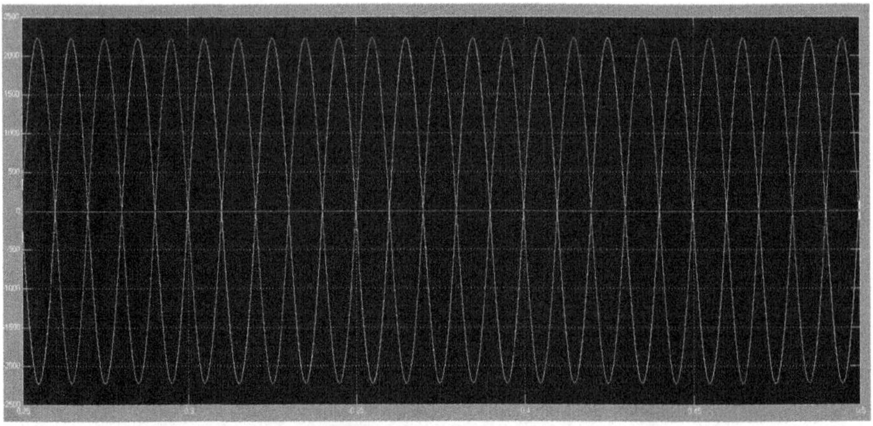

Figure 7.4: DVR output 3 phase line

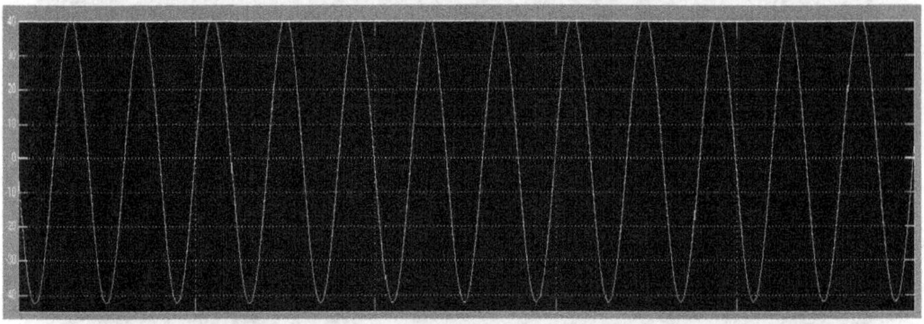

Figure 7.5: bus output voltage whit DVR

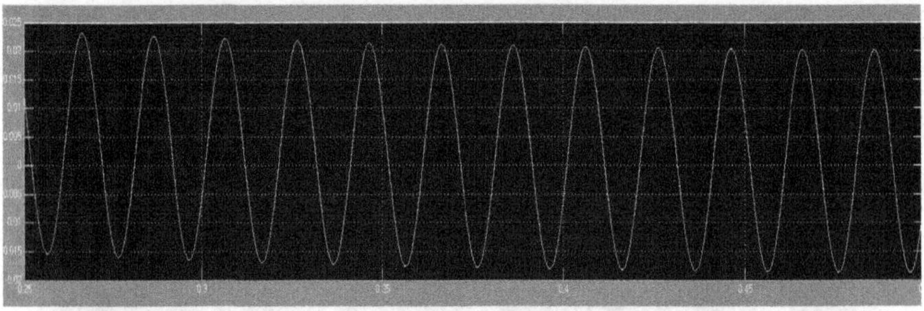

Figure 7.6: Bus output voltage whitout DVR

*****

CHAPTER 8
CONCLUSION

8.1 Introduction

Based on this project, one of the methods to mitigate the voltage sag is by using dynamic voltage restorer system. In order to investigate whether the DVR is able to deal with this problem, Matlab Simulink was selected in order to simulate the system and mitigate the voltage sag. Based on the simulation that had been done, it can be proved that DVR is the dynamic fast response devices that able to overcome the devices. Even though the system cannot compensate 100 percent of voltage during sag, it is an acceptable because the output voltage after compensation still in range of the nominal value. The simulation was implemented by using the distribution network where the effectiveness of the DVR system is better compared to the transmission network. Since DVR is the custom power devices, there were so various combinations of main component that can be combined in order to get better results. For example, using GTO as a switching device, using SVPWM as the device for generating gating signal or maybe can use the other type of filter in order to eliminate the harmonics In this chapter, reviews of all the major research literatures, presented in the area of power quality improvements, are given. Further, power quality issues like voltage sag, swell and harmonics are also explained. A brief description of the impact of power quality and its standards, along with solutions for power quality improvements are highlighted, in Table 2.1. It is clear that majority of the research works focus on voltage sag, harmonics and swell. However, the existing compensation devices have not focused on long voltage interruption problem and also the utilization of photovoltaic energy source. The compensating system with DVR does not compensate sustained interruption of power supply, rather it compensates short interruptions. The usage of DSTATCOM cannot supply active power to the system. The proposed PV-UPQC system mitigates the power quality problems and thereby, provides effective solutions for the power quality issues. Therefore, in this chapter, this research gap in the power quality improvement is highlighted. Potential mitigation of long voltage interruption and the provision of active power supply to the system are the major requirements. Consequently, there is a huge requirement for fast response systems compared with the traditional methods. The proposed PV-UPQC system improves considerably the power quality with respect to voltage sag, swell, harmonics and long voltage interruption.

A. Advantages

- DVR is mostly preferred because it is less¬ expensive. If compared to DSTATCOM and UPS, DVR is a lot cheaper.
- UPS requires a lot of maintenance which may¬ be due to problems of battery leakage and replacement. DVR is small in size and is a better power¬ effective device as compared to others link UPS, SMES and DSTATCOM.
- SVC cannot control active power flow that¬ creates a reason of preference for DVR in spite of the fact that SVC is better than DVR.

B. Limitations

- DVR has a limited current conduction and¬ voltage injection capability which is due to such design of DVR to keep its cost low as well as to reduce the voltage drop across it in standby mode.
- In order to reduce the cost of DVR, the energy¬ storage size of DVR is kept low. Due to voltage dips, the stored energy can deplete fast and therefore to avoid load tripping due to insufficient stored energy, an adequate control is required.

8.2 MODELLING AND SIMULATION

This paper Compensation Voltage Sag using Dynamic voltage restorers (DVR) The power and voltage quality problems such as voltage sage, swells and others. Also an overview of dynamic voltage restorer (DVR) is presented. DVRs are effective recent custom power devices for voltage sags and swells compensation. They inject the appropriate voltage component to correct rapidly any anomaly in the supply voltage to keep the load voltage balanced and constant at the nominal value. The Dynamic Voltage Restorer (DVR) is considered to be an efficient solution due to its relatively low cost and small size; also it has a fast dynamic response show in fig 7.2 to 7.6

REFERENCES

[1] A. Ghosh and G. Ledwich, "Compensation of distribution system voltage using dvr," IEEE Transactions on Power Delivery, vol. 17, no. 4, pp. 1030–1036, Oct. 2002. 188

[2] A. Ghosh and G. Ledwich, "Structures and control of a dynamic voltage regulator (dvr)," in IEEE Power Engineering Society Winter Meeting, vol. 3. IEEE, 2001, pp. 1027–1032.

[3] A. Ghosh, A. Jindal, and A. Joshi, "Design of a capacitor-supported dynamic voltage restorer (dvr) for unbalanced and distorted loads," IEEE Transactions on Power Delivery, vol. 19, no. 1, pp. 405–413, Jan. 2004.

[4] S. Sasitharan and Mahesh K. Mishra, "Constant switching frequency band controller for dynamic voltage restorer," IET Power Electronics, vol. 3, no. 5, pp. 657–667, Sept. 2010.

[5] S. Sasitharan, Mahesh K. Mishra, B. Kalyan Kumar, and V. Jayashankar, "Rating and design issues of dvr injection

[6] M. H. J. Bollen, "Understanding power quality problems: Voltage sags and interruptions," ser. Power Engineering, Piscataway, NJ: IEEE Press 2000.

[7] R. C. Dugan, M. F. McGranaghan, and H. W. Beaty, Electrical Power System Quality, New York: McGraw-Hill, 1996.M. T.Neela Srividya et al, International Journal of Computer Science and Mobile Computing, Vol.3 Issue.10, October- 2014, pg. 380-386 © 2014, IJCSMC All Rights Reserved 386

[8] M. T. Aung, J. V. Milanovic, and C. P. Gupta, "Propagation of assymmetrical sags and the innfluence of boundary crossing lines on voltage sag prediction," IEEE Trans. Power Del., vol. 19, no. 4, pp. 1819-1827, Oct 2004.

[9] I. Hunter, "Power quality issues: A distribution company perpective," Power Engineering Journal, vol. 15, pp. 75-80, 2001.

[10] J. V. Milanovic and C. P. Gupta, "Probabilistic of financial losses due to interruptions and voltage sags–Part I: The methodology," IEEE Trans Power Del., vol. 21, no. 2, pp. 918-924, 2006.

[11] J. V. Milanovic and C. P. Gupta, "Probabilistic of financial losses due to interruptions and voltage sags–Part II: Practical implementation," IEEE Trans Power Del., vol. 21, no. 2, pp. 925- 932, 2006.

[12] A. Ghosh and Gerard, "Power quality enhancement using custom power devices," Kluwer Power Electronics and Power System Series, 2002.

[13] A. Ghosh and G. Ledwich, "Compensation of distribution system using DVR," IEEE Trans. on Power Del., vol. 17, pp. 1030-1036, 2002.

[14] D. V. Hertem, M. Didden, J. Driesen, and R. Belmans, "Choosing the correct mitigation method against voltage dips and interruption: A customer-based approach," IEEE Trans. on Power Delivery, vol. 22, no. 1, pp. 331-391, Jan 2007.

[15] S. S. Choi, J. D. Li and D. M. Vilathgamuwa, "Dynamic voltage restorating with minimum energy injection," IEEE Trans. On Power Sys., vol. 15, pp. 51-57, Feb 2000.

SECTION - C
SIMULATION OF LIGHTNING ARRESTOR AND STUDY OF OVERVOLTAGES UNDER DIFFERENT FAULT CONDITIONS

CHAPTER 1
INTRODUCTION

1.1 Introduction

Lightning is a sudden electrostatic discharge that occurs during a thunder storm.

This discharge occurs between electrically charged regions of a cloud between two clouds or between a cloud and the ground .the charged regions in the atmosphere temporarily equalize themselves through this discharge referred to as a strike if it an hits an object on the ground and a flush if it occurs within a cloud. Lightning causes light in the form of plasma and sound in the form of thunder. Lightning may be seen and not heard when it occurs at a distance too great for the sound to carry as for as the light from the strike or flash.

Lighting is a gaint spark of electricity in the atmosphere between clouds, the air or the ground in the early stages of development air acts as an insulator between the positive and negative charges in the cloud and between the cloud and the ground .when the opposite charges builds up enough this insulating capacity of the air breaks down and there is a rapid discharge of electricity that we know as lightning. The flash of lightning temporarily equalizes the charged regions in the atmosphere until the opposite charges build up again.

1.2 Lightning Phenomerna

1.2.1 Mechanism of Lightning

Benjamin Franklin performed his famous experiment of flying kite in thunder cloud in 1745.before his discovery the lightning used to be considered to be "act of god" He also proved that the lightning stroke is because of the discharge of electricity. He also invented lighting rods for fixation on tall buildings and grounding for their protection from lightning strokes. Thus Benjamin Franklin is a pioneer scientist in this field.

Lightning is a huge spark, which is due to the electrical discharge taking place between the clouds, within the same clouds and between the clouds and earth. Large numbers of discharges occur between or within clouds than to earth, and enough of them terminate on to the earth and result in serious hazards.

Lightning discharge requires the puncture of air between the cloud and the earth.

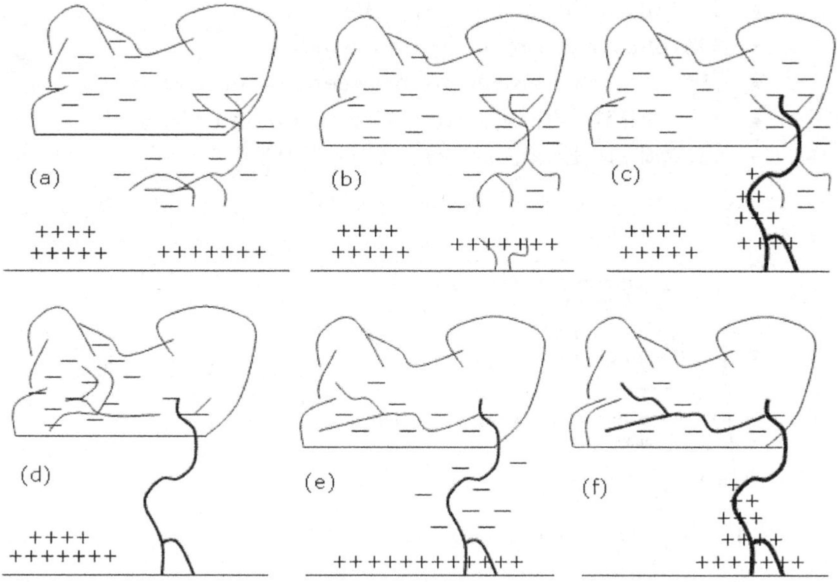

Figure 1: Diagram showing various stages of lighting stroke between cloud and ground

The mechanism of lighting discharge can we well explained with the help of fig. 1. Whenever A local charge concentration near a cloud causes the potential gradient to exceed the critical breakdown value, the air surrounding gets ionized and an electric streamer starts from the cloud toward the earth. The current in the streamer is of the order of 100 A and the speed of the Streamer is 0.15 m/μs. this streamer is known as pilot streamer because it leads to the Lightning phenomenon.

1.2.2 Waveshape of Lightning Stroke

Surge caused due to lightning discharge are generation Non-oscillatory nature with a very steep wave front and have gradually decreasing tail. The Wave is designated by (i) maximum or crest value (ii) the front time t_1 which is the time to due To rise to the peak value and (iii) the duration t_2 which is the time from 0 to the point where The voltage or current has decayed to one half its peak or crest magnitude. It has been observed that:

1. t_1 varies from 1μs to 10 μs
2. t_2 varies from 10μs to 100μs
3. Maximum value of peak current is 400 kA
   - 2.5% of all the strokes are below 10kA
   - 84% of all the strokes are below 50kA
   - 11% of all the strokes are between 50 and 100kA
   - 2% of all the strokes are between 100 and 150kA
   - 0.5% of all the strokes are above 150A

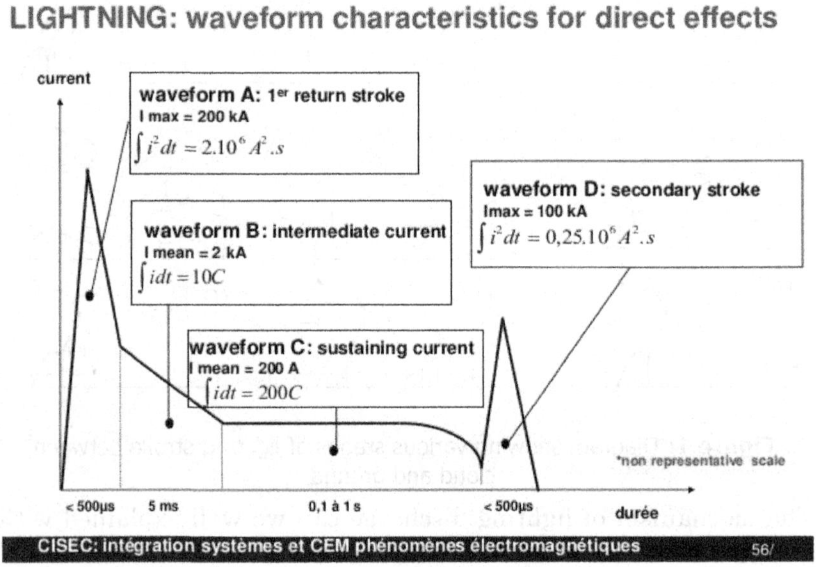

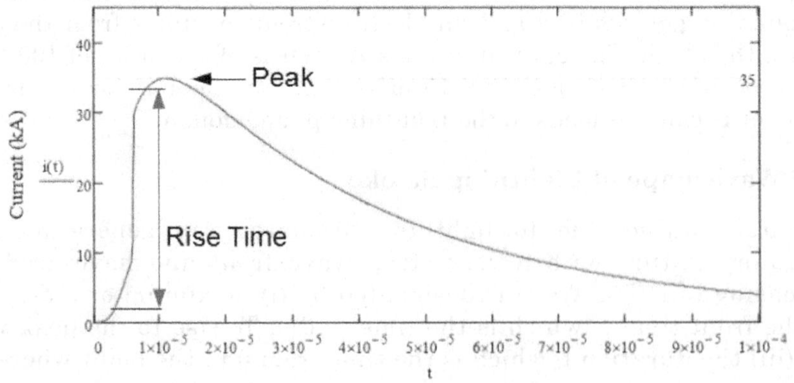

1.2.3 Lightning Strokes

There are two main ways in which lightning affects a line: by a direct stroke, and by electrostatic induction.

1. DIRECT LIGHTNING STROKES:

A direct stroke can take place in several ways. In one way a cloud attains a large amount of charge, it induces an opposite charge on the tall objects, such as temples, churches or mosques. In between these two charges are electrostatic field is set up and when the intensity of electrostatic field become sufficiently great to ionize the neighboring air, the air breaks down and discharge takes place between the cloud and the object. Such a discharge is known as a stroke, and is characterized by the comparatively long time taken to produce it and the fact that it strikes the highest and most sharply pointed object in the neighborhood. Another way results in a much more sudden stroke, which is produced in the manner. Three clouds are involved. Clouds 1 and 3 are positively charged and cloud 2 is negatively charged. The potential of cloud 3 is reduced due to presence of the charged cloud 2. On flash over from cloud 1 to cloud 2, both these clouds are discharged rapidly and cloud 3 assumes a much higher potential and flashes to earth very rapidly. This is the B stroke, and is characterized by its rapidity and the fact that it ignores tall objects and reaches ground in the random manner. Thus the stroke B is the result of stroke A between cloud 1 and 2. It is the most dangerous stroke. No protection can be designed to avoid stroke B where as protection can be provided against stroke A by channelizing the charge through a lightning conductor placed at the highest point on the building and the earthling properly at the other hand. The stoke B is sometime called the induced stoke

Lightning can have **direct effects** but **indirect effects as well,** depending on whether it strikes a structure directly or not:

Fig. 1

Fig. 2

Fig. 3

Fig. 4

Fig. 1. Lightning striking a structure directly.
Fig. 2. Lightning striking the ground near a structure.
Fig. 3. Lightning striking services connected to a structure.
Fig. 4. Lightning striking the ground near services connected to a structure.

2. SURGE DUE TO ELECTOSTATIC INDUCTION:

The most of the surge due to transmission system are due to lightning and are cause by electrostatic induction in the manner shown in figure .A cloud electrostatic charged and lying above a transmission line will induced in the adjacent section of the a corresponding charge of opposite sigh known as the bound charge. This charge has the maximum value below the cloud and then gradually tails away. At the distant end of the line, charges of like sign to that in the cloud are located. These charges on the line will not flow since it is a bound charge. The positive charges on the far end of the line will however leak to earth slowly through insulated metallic part etc, thus leaving only the negative charge on the line.

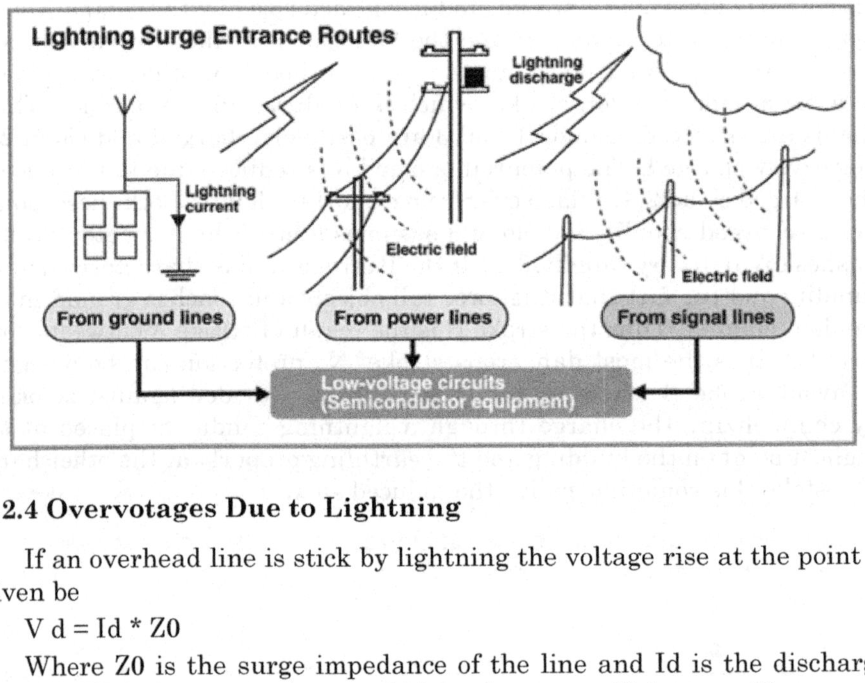

1.2.4 Overvotages Due to Lightning

If an overhead line is stick by lightning the voltage rise at the point is given be

V d = Id * Z0

Where Z0 is the surge impedance of the line and Id is the discharge current. The above equation would have been true if the travelling waves flow

Only in one direction,

But since they travel in both directions, the current is halved and over voltage is given by the expression.

V d = ½ Id* Z0

A direct stroke may cause a potential of many million volts. For example if Id is taken 30kA and Z0 = 500Ω, then

V d = ½ * 30 * 1000 * 500 = 7.5 *10^6 v

A direct stroke on the tower or earth conductor causes the voltage rise of the latter to a value given by the expression.

$V d = Id * RE + L (di/dt)$

Where Id is discharge current, RE is the tower Footing resistance, L is the inductance of the tower and di/dt is the slope of current wave.

For EHT towers of normal height, L is taken equal to 10μH

Taking RE = 5Ω, Id = 30kA & di/dt = 10kA /μs, we have

$V d = 30 * 10^3 * 5 + 10 * 10^{-6} * ((10*1000)/(1*10^{-6})) = 250$ kV

For all purpose only travelling waves through the lines are considered for protection.

1.2.5 Protection Against Lightning

All the electrical equipment must be protected from severe damage due to the lightning strokes.

The problem of protection of power system against lightning can be studied under the following Heads.

1. Protection of power stations and substations from direct lightning strokes.
2. Protection of overhead transmission lines from direct lightning strokes.
3. Protection of electrical equipment from travelling waves.

1.2.6 Insulation Coordination

The selection of suitable values for the insulation levels 0f the various components in any

Electrical system and their arrangement in a rational manner is called insulation co-ordination.

The insulation level of an apparatus is defined as that combination of voltage values which characterize

Its insulation with regard to its insulation with regard to its capability of withstanding the dielectric stresses. A proper co-ordination of insulation seeks to ensure:

1. that the insulation provided shall withstand all normal working stresses and a majority of abnormal ones
2. the efficient discharge of over voltages due to internal and external causes
3. that the breakdown shall occur only due to external flashover
4. that the positions at which breakdown occurs shall be those where breakdown may cause no or comparatively littlies damage.

The insulation problem in a power system shall, therefore, involve the following:

1. Determination of line insulation.
2. Selection of basic impulse insulation level and insulation levels of other equipment.
3. Selection of lightning arresters.

1.2.7 Lightning Arrester Selection

For simplify, the process of making an arrester application may be reduced to seven steps.

1. To determine the magnitude of the power frequency phase to ground voltage expected at the proposed arrester location during phase to ground fault, or other abnormal conditions which cause higher voltages to ground than normal.
2. To make a tentative selection of the power frequency voltage rating of the arrester.
3. To select the impulse current likely to be discharge through the arrester.
4. To determine the maximum arrester discharge voltage for impulse current & type of arrester selected.
5. To establish the full-wave impulse voltage with stand the level of the equipment to be protected.
6. To make certain that the maximum arrester discharge voltage is below the full wave impulse; withstand level of the equipment insulation to be protected, by adequate margin.
7. To establish the separation limit between the arrester & the equipment to be protected.

1.3 LIGHTNING ARRESTER

A protective device for electrical equipment that reduces excessive voltage resulting from lightning to a safe level by grounding the discharge.

A lightning arrester is a device used on electrical power system and telecommunications systems to protect the insulation and conductors of the system to protect the insulation and conductors of the system from the damaging effects of lightning. The typical lightning arrester has a high voltage terminal and a ground terminal. When a lightning surge travels along the power line to the arrest, the current from the surge is diverted through the arrestor, in most cases to earth.

In telegraphy and telephony, a lightning arrestor is placed where wires enter a structure, preventing damage to electronic instruments within and ensuring the safety of individuals near them. Smaller versions of lightning arrest, also called surge protectors, are devices that are connected between each electrical conductor in power and communication system and the earth. These prevent the flow of the normal power or signal current to ground, but provide a path over which high voltage lightning current flows, by passing the connected equipment. Their purpose is to limit the rise in voltage when a communication or power line is struck by lightning or is near to a lightning strike.

If a protection fails or is absent, lightning that strike the electrical system introduces thousands of kilovolts that may the transmission lines, and can also cause several damage to transformers and other electrical and electronic devices.

1.3.1 Types of Lightning Arrester

The lightning arrester protects the electrical equipment from lighting .it is placed very near to the equipment and when the lightning occurs the arrester diverts the high voltage wave of lightning to the ground. The selection of arrester depends on the various factors like voltage, current, reliability, etc. the lightning arrestor is mainly classified into twelve types.

These types are;

1. Road gap arrester
2. Sphere gap arrester
3. Horn gap arrester
4. Multiple-gap arrester
5. Impulse protective gap
6. Electrolytic arrester
7. Expulsion type lightning arrester
8. Valve type lightning arrester
9. Auto valve arrest
10. Oxide film arrester
11. Thyrite lightning arrester
12. Metal oxide lightning arresters

1.3.2 Metal Oxide Lightning Arresters

A new type of lightning arrester element has been developed and marketed since seventies. This arrester has its element with non linear resistance characteristics and the main constituent of the arrester element being zinc oxide; it is economical as compared to thyrite arrester. The degree of non-linearity of this arrester element is much more than that in the thyrite arrester. No series gap is required in the arrester.

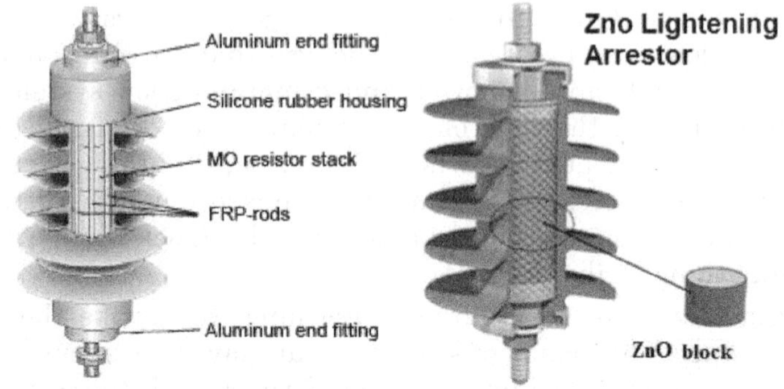

This ZNO arrester is of simple construction, better surge protection and more stable protection characteristics. Construction wise, it is almost similar to the valve type arrester but the valve element is a dense ceramic body made up of ZNO and some other metallic oxides.

Such an arrester is simple in construction and has superior energy absorption capability, better surge protection, more stable protective characteristics and substantial reduction in overvoltage across the equipment as compared to thyrite arrester.

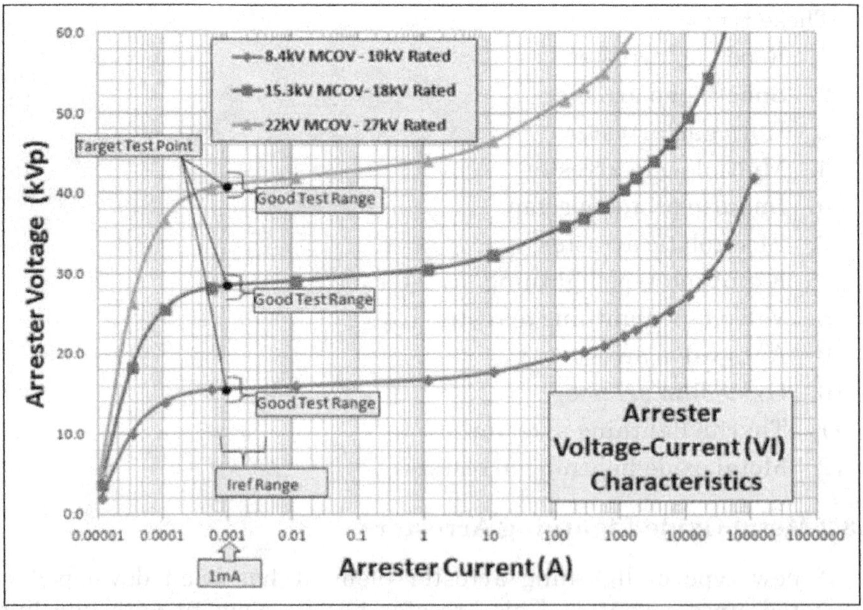

The zinc oxide is a semiconducting material of N-type. It is pulverised and finely grained. More than ten doping materials are added in the form of fine powders of insulating oxides such as Bismuth (Bi_2O_3), Antimony Trioxide (Sb_2O_3), Cobalt Oxide(CoO), Manganese Oxide (MnO_2), Chromium oxide (Cr_2O_3). The powder is treated with some processes, and the mixture is spray dried to obtain a dry powder.

The dry powder is compressed into disc-shaped blocks. The blocks are sintered to obtain a dense poly- crystalline ceramic. The metal oxide resistor disc is coated with a conducting compound to protect the disc from undesirable environmental effect.

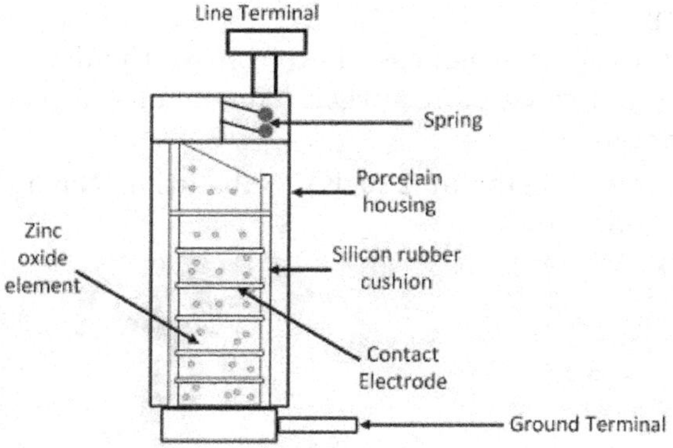

Zinc Oxide Surge Arrester

The conducting coating also provides proper contacts and uniform current distribution. The disc then enclosed in a porcelain housing filled with nitrogen gas or SF_6 gas. Silicon rubber is used to keep the disc in a position. It also helps in heat transfer from disc to the porcelain housing. The disc is held under pressure using suitable springs.

The ZnO element eliminates series sparks gaps in the diverter. The voltage drop in ZnO diverter takes place at the grain boundaries. There is a potential barrier at the boundary of the each grain of ZnO and this potential barrier control the flow of current from one grain to the next.

At normal voltage, the potential barrier does not allow the current to flow through it. At over voltage the barrier collapse and sharp transition of current from insulating to conducting state take place. The current start flowing and the surge is diverted to ground.

After the travelling of the surge, the voltage across the diverters falls, and the current is reduced to the negligible value of the resistor units, and there is no power follow current.

Advantages of Metal Oxide Surge Arrester

The metal oxide surge arrester has the following merits:

- It eliminates the risk of spark over and also the risk of shock to the system when the gaps break down.

- It eliminates the need of voltage grading system.
- At the normal operating condition, the leakage current in the ZnO is very low as compared to other diverters.
- There is no power follow current in ZnO diverter.
- It has high energy absorbing capability.
- ZnO diverters possess high stability during and after prolonged discharge.
- In ZnO diverter, it is possible to control the dynamic overvoltages in addition to switching surges. This results in economic insulation coordination.
- Lightning arrestor at 220 KV Substation Nayagaon Jabalpur is shown in figure

Power Diode and Its Characteristics

Power semiconductor diode is the "power level" counter part of the "low power signal diodes" with which most of us have some degree of familiarity. These power devices, however, are required to carry up to several KA of current under forward bias condition and block up to several KV under reverse biased condition. These extreme requirements call for important structural changes in a power diode which significantly affect their operating characteristics. These structural modifications are generic in the sense that the same basic modifications are applied to all other low power semiconductor devices (all of which have one or more p-n junctions) to scale up their power capabilities. It is, therefore, important to understand the nature and implication of these modifications in relation to the simplest of the power devices, i.e., a power semiconductor diode.

Power Diodes of largest power rating are required to conduct several kilo amps of current in the forward direction with very little power loss while blocking several kilo volts in the reverse direction. Large blocking voltage requires wide depletion layer in order to restrict the maximum electric field strength below the "impact ionization" level. Space charge density in the depletion layer should also be low in order to yield a wide depletion layer for a given maximum Electric fields strength. These two requirements will be satisfied in a lightly doped p-n junction diode of sufficient width to accommodate the required depletion layer. Such a construction, however, will result in a device with high resistively in the forward direction. Consequently, the power loss at the required rated current will be unacceptably high. On the other hand if forward resistance (and hence power loss) is reduced by increasing the doping level, reverse break down voltage will reduce. This apparent contradiction in the requirements of a power diode is resolved by introducing a lightly doped "drift layer" of required thickness between two heavily doped p and n layers as shown in Fig 2.3(c). Fig 2.3 (a) and (b) shows the circuit symbol and the photograph of a typical power diode respectively.

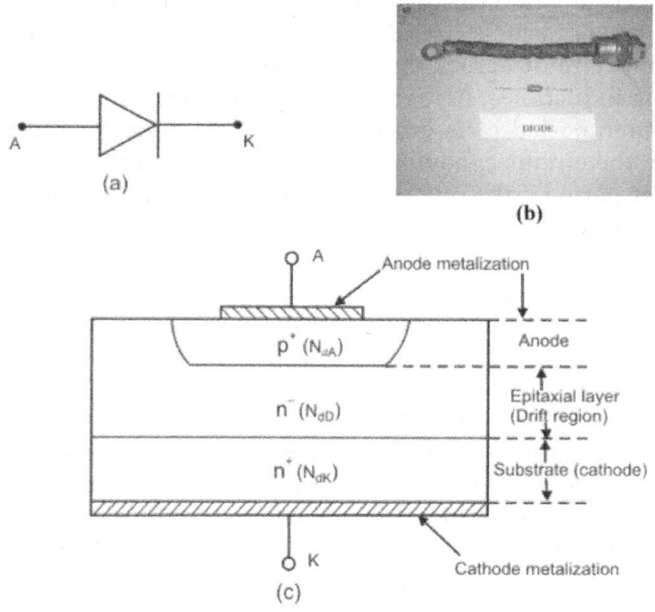

Figure 2.3: Diagram of a power: (a) circuit symbol, (b) photograph, (c) schematic cross section

Switching Characteristics of Power Diodes Power Diodes take finite time to make transition from reverse bias to forward bias condition (switch ON) and vice versa (switch OFF). Behaviour of the diode current and voltage during these switching periods are important due to the following reasons. • Severe over voltage / over current may be caused by a diode switching at different points in the circuit using the diode. • Voltage and current exist simultaneously during switching operation of a diode. Therefore, every switching of the diode is associated with some energy loss. At high switching frequency this may contribute significantly to the overall power loss in the diode. Observed Turn ON behaviour of a power Diode: Diodes are often used in circuits with di/dt limiting inductors. The rate of rise of the forward current through the diode during Turn ON has significant effect on the forward voltage drop characteristics

Salient features of this characteristics are: • The diode current does not stop at zero, instead it grows in the negative direction to Irr called "peak reverse recovery current" which can be comparable to IF. In many power electronic circuits (e.g. choppers, inverters) this reverse current flows through the main power switch in addition to the load current. Therefore, this reverse recovery current has to be accounted for while selecting the main switch. • Voltage drop across the diode does not change appreciably from its steady state value till the diode current reaches reverse recovery level. In many power electric circuits (choppers, inverters) this may create an effective short circuit across the supply, current being limited only by the stray wiring inductance. Also in high frequency switching circuits (e.g, SMPS) if the time period t4 is comparable to switching cycle qualitative modification to the circuit behaviour is possible. • Towards the end of the reverse recovery period if the reverse current falls too sharply, (low value of S), stray circuit inductance may cause dangerous over voltage (Vrr) across the device. It may be required to protect the diode using an RC snubber. During the period t5 large current and voltage exist simultaneously in the device. At high switching frequency this may result in considerable increase in the total power loss. Important parameters defining the turn off characteristics are, peak reverse recovery current (Irr), reverse recovery time (trr), reverse recovery charge (Qrr) and the snappiness factor S. Of these parameters, the snappiness factor S depends mainly on the construction of the diode (e.g. drift region width, doping lever, carrier life time etc.). Other parameters are interrelated and also depend on S. Manufacturers usually specify these parameters as functions of diF/dt for different values of IF. Both Irr and Qrr increases with IF and diF/dt while trr increases with IF and decreases with diF/dt.

*****

CHAPTER 2
REVIEW OF LITERATURE

Overvoltages occurring in electrical power networks are caused by changes of power network operation conditions (e.g.faults, lightning strikes or switching operations). In order to limitation of overvoltages, Metal Oxide Surge Arresters (MOSA)are installed in electrical networks. To analyze and determine overvoltages values in electrical power systems, dedicated simulations are necessary. Increase voltage over the nominal value is very dangerous for insulating system of electrical devices installed in electrical power network. These phenomena are able to damage electrical equipment installed in power network or even may cause power outage. For purpose reduce overvoltages, metal oxide surge arresters are installed in electrical network. Performing simulations by means of specialized software is necessary to determine overvoltages values occurring in electrical network during transient states. Thus, for this purpose, equivalent mathematical models of surge arresters are indispensable. Frequency-dependent models provide more accurate representation of real surge arresters for transient states analyses. The effect of overvolages on lightning arrestor, Modelling, reliability assessment, fault patterns and efficiency of surge arrestor is available in literature. Some of the recent researches are reviewed below:-

Leela A M, K N Ravi et al (2017) in the paper "Simulation and Analysis of ZnO block Characteristics Using MATLAB" (IEEE INTERNATIONAL CONFERENCE ON INTELLIGENT TECHNIQUES IN CONTROL, OPTIMIZATION AND SIGNAL PROCESSING) mentioned that Lightning arresters are vital protective components in power system. Ageing of the arrester is an important issue under research. Simulation of the arrester and analysis using the simulation tool plays an important role in validating the experimental results. The leakage current that flows through the arrester blocks due to continuous electric stress on the arrester, ambient temperature, pollution on the surface of the arrester etc., causes heating of the arrester blocks. The increase in temperature of the arrester blocks causes the blocks to age. In order to understand the various mechanism of ageing it becomes necessary conduct experiments on the arrester subjecting it to various conditions and analyzing. It has been observed by many researchers that the resistive component of the leakage current and in particular the third harmonic component of the resistive component of the total leakage current is responsible for heating of the blocks. In this paper ZnO block of the surge arrester is simulated using MATLAB. V-I characteristics with resistive component of the total leakage current at different temperatures obtained. The resistive and capacitive components of the total current at different temperatures obtained. Also the harmonic components of the resistive component of the total leakage current obtained and analyzed. From the analysis of results of simulations it is observed that they conform to the experimental results of the earlier researchers. From the waveforms of Total Leakage Current and its Resistive and Capacitive Components, it is observed that the resistive component dominates as the voltage is increased and also as the temperature increases. This indicates that the root cause of thermal ageing of the arrester block i.e., the resistive component of the leakage current depends on the temperature and voltage. From the results of the simulation carried out it can be concluded that simulation using MATLAB gives satisfactory results for the arrester block. The simulation can be extended for the arrester as a whole including the housing. Also it can be extended for simulating the pollution layer on the housing and the coupling of the surface current that flows on the polluted layer with the inside blocks and further contributes to ageing of the arrester.

Valdemir S. Brito, George R. S. Lira, (2016), in the paper "A Wide-Range Model for Metal-Oxide Surge Arrester" (IEEE TRANSACTIONS ON POWER DELIVERY) in this paper authors present an electric model of metal-oxide surge arrester (MOSA). The proposed electric model accurately represents the MOSA in a wide range of frequencies and amplitudes. The model was developed and validated based on MOSA electrical behaviour in each one of the three operating regions of the zinc-oxide (ZnO) surge arresters, and in a database composed of voltage and current waveforms measured from tests performed in twelve ZnO varistors having different physical dimensions and electrical characteristic - from five different manufacturers. These varistors were subjected to different voltage levels in the low current region, and multilevel amplitude of switching current impulses (30/60 μs), lightning current impulses (8/20 μs), high current impulses (4/10 μs) and fast-front current impulses (1.5/26 μs and 3/6 μs), encompassing the three regions of operation and a wide-range of frequencies and amplitudes. The results provided by the MOSA wide-range (MWR) model were compared with that obtained in laboratory. The MWR model has shown good agreement in terms of waveform, peak value, and absorbed energy for the evaluated cases.In this paper, MOSA wide-range model (MWR model) was presented. This model is based on MOSA electric behaviour in its three operating regions. The model was validated from measurement of voltage and current in MOSA regions of operation. The model is simple, and all its parameters are easily found in the manufacturer's datasheet. The different studies with MOSA can be performed using only the MWR model. This avoids error of representation when the studies involves more of one region, as in the case of the load-shedding due to a short-circuit. Furthermore, the MWR model is easy to use and allows the permanent representation of the MOSA in the simulated electrical system. From the measurement results, significant differences in the behaviour of the V -I curve, in the high nonlinearity region, in accordance with the line discharge class of the MOSA were verified. The waveforms of the voltages or currents supplied by the MWR model were similar to the lab tests results. The differences observed in the high nonlinearity and high current regions can be attributed, at this moment, to the necessity of improving the inductive coupling compensation method. The method provided in the standards for very fast transients, apparently, is not sufficient. From the measurement results, it has been verified that the inductive coupling present in the MOSA residual voltage measurement is significant, too, to transients such as switching current impulses (30/60 μs), contrary to recent technical standards that claim to be needing some kind of compensation for very fast transients (front times equal or less than 1 μs).

Fernando Gontijo Bernardes, Daniel Carrijo Polonio Araujo and Rodney R. Saldanha (2017) Automatic detection of fault patterns in lightning arresters, IEEE International Symposium on Lightning Protection observed that Lightning arresters are equipments with a fundamental role in the protection of surges in substations. Thermographic images have been used in large scale to prevent faults in these devices. This paper presents a new framework aimed to nondestructive inspection of surges and medium voltage equipments. It applies a fuzzy clustering algorithm to automatic identify fault patterns in thermographs of medium voltage equipments, minimizing the needs of qualified manpower. The technic can also be used in thermography drones in a way to acquire the images remotely and automatic processes the fault triggering the staff only when a fault is detected. Experimental results prove the effectiveness of the method identifying hot spots in the devices. Thermography is a current, inexpensive, simple, noninvasive solution for analyzing faults or inappropriate behaviour in electrical equipment that generates heat such as lightning arrester and electrical circuits. It is a tendency to reduce the human interaction in the thermographic analysis in order to avoid the bias of the solution pointed out by the specialist, reducing costs and gaining scale using automatic detection systems. The fuzzy C-Means (FCM) together with subtractive clustering were used in works with satisfactory results in the identification of patterns in thermographic images. The developed system correctly classifies thermographic images of lightning arresters and medium voltage equipment. It brings a contribution to the literature specifying a way to reduce the human interaction in the fault diagnosis of lightning arresters. Also improves the accuracy found in similar problems and, in addition, can be used in the classification of thermographs toke by remote devices like drones. Reducing maintenance costs of high voltage equipments and providing a way to explore the new technologies. In future works the system can be improved including new variable of analyses, making the system even more precise. Also, others fault patters could be included adapting to new failure modes.

Nikolov, Dimitrova, Georgiev and Vasileva (2017) in paper Reliability Assessment of Electricity Distribution Substation Surge Protection System SPS [IEEE International Spring Seminar on Electronics Technology] in this article concerns reliability assessment of an electricity distribution substation surge protection system, composed of surge arresters. For this purpose a reliability analysis is performed and also a surge protection system reliability mathematical model is suggested. Based on it mathematical formulae are worked out intended for practical calculation of proper reliability indices values. The basic dependences derived are graph Based on calculations performed in regard to probability for SPS reliable operation it can be concluded that the increase of SA failure rates, caused by the increase of the power dissipated by the respective SA, or by a large voltage loading, leads to decrease of the probability for SPS reliabile operation. The worst case during SPS operation obviously is this, when a lightning strokes on incoming OHL or on busbar system. During the lightning impact the voltage and temperature stress of the SA, respectively of SPS, are extreme. From Figure 2 and Figure 3 is visible that the increase of the coefficient values of π T and π S causes a total decrease of the probability for SPS reliability operation. It is also visible that the values of π S are more crucial for SPS reliable operation. The SPS reliability model suggested can also be applied in cases when the maintenance engineers are trying to select a proper set of SA in order to ensure as high as possible level of the substation equipment surge protection and SPS reliable operation, considering the SA types at the disposal. This surge protection subsystem is the subject of the analysis in this paper. The authors studied SPS most frequently, analyze the surge protection peculiarity during a lightning stroke at different values of the Earth conductivity. Sometimes they bring into use a coefficient of surge protection in order to determine the effectiveness of surge protection devices allocation. There are also exist approaches to assessment of the probability danger overvoltage wave affecting the equipment to arise from a lightning stroke . During this study in the area of surge protection against atmospheric overvoltages, affecting a SPS composed by metal-oxide SA, it is not ascertained there existence of any approaches and/or models allowing the reliability assessment of SPS. As the SPS performs a very important system function, its reliable operation is crucial. This is thereason why the assessment of SPS reliability is so necessary. Derivation of a mathematical reliability model of SPS composed by SA, allowing reliability assessment of such system is the object of the study.

L. Liljestrand and E. Lindell (2015) "Efficiency of surge arresters as protective devices against circuit breaker induced overvoltages" IEEE Transactions on Power Delivery, In this paper authors studied the possible occurrence of transient overvoltages depends on the interaction between different components in a power system. In a cable system, circuit breaker operations may cause overvoltages, depending on the system configuration and type of circuit breaker operations performed. Protective devices may be required during certain conditions. Surge arresters are a well-known product that could be applied. Surge arresters were originally developed as protection against lightning induced overvoltages on overhead lines. However, they can also be used for protection against vacuum circuit breaker induced transients in cable systems, and this is what the current publication reports on. Different surge arrester configurations were evaluated and two different surge arrester states are discussed; a "high current state" when the circuit breaker is closed and a high current passes through the surge arrester and a "low current state" when the circuit breaker is open and only a low current passes through the surge arrester. Finally, a new surge arrester configuration was evaluated and found to be advantageous in preventing amplification of internal resonances in the transformer.There are four phenomena related to vacuum circuit breaker induced transient overvoltages when energizing and de-energizing an inductive load: 1. Multiple pre-strikes during energizing 2. Current chopping during de-energizing 3. Multiple reignitions during de-energizing 4. Virtual current chopping during de-energizing. In general, energizing does not cause very high transient overvoltages since the voltage is limited by the reduced electrical withstand strength of the closing contacts even though multiple prestrikes could occur when closing the circuit breaker at an inductive load. Current chopping during de-energizing could cause overvoltages and initiate multiple re-ignitions which could cause voltage escalations. It was found during the testing that the highest voltages were recorded when the multiple re-ignitions caused virtual current chopping. Surge arresters are an efficient type of protective device for limitation of transient overvoltage amplitudes. The voltage will be limited to a well-defined level. When used for limitation of circuit breaker induced transients, two different situations need to be distinguished. The first one is valid when the circuit breaker is conducting and the surge arrester limits overvoltages due to prestrikes or reignitions. In this case, a high current, in the range of kiloamps, will pass through the arrester and the voltage will be limited at about 80 % to 100 % of the protective level. When the circuit breaker is open, however, the surge arrester provides an optional return current path for the magnetically stored energy in the winding, similar to a freewheeling diode. In this case, the maximum current will be equal to the virtual current chopping level, which was the peak load current in to order of 10 A in the testing. In this case, the voltage will be limited at a significantly lower amplitude; about 70 % of the protective level. When the surge arrester limits the overvoltage on the load side of the circuit breaker, the voltage across the circuit breaker, i.e. the recovery voltage, will also be limited. The result of this is that the reignitions will stop; which is an important conclusion. Typical values of the nominal current in practical installations may range between 50 A and 500 A. Still, since most medium voltage transformers (especially dry type) have limited power, the chopped current will in most cases be well below 10 kA where the protective level of the surge arrester is defined. This means that in most cases the surge arrester will limit the voltage to a lower value than might originally have been expected. Installing surge arresters only phase-to-ground, effectively two surge arresters in series, will limit the voltage across the winding. Installing surge arresters phase-to-phase in addition will limit the voltage across the winding to half the amplitude and is therefore recommended. In this configuration, basically no current passes through the surge arresters to ground, but they are anyhow not recommended to be removed, since that may cause a significant DC offset in the terminal voltages. An innovative configuration where the surge arrester across the winding is "split" into two parts and the midpoint is connected to the midpoint of the transformer winding was also tested. This configuration limits the voltages across the respective halves of the transformer winding even further and limits excitation of internal resonances. Transformers equipped which such a protection configuration are called Transient Voltage Resistant Transformers (TVRT). The overvoltage amplitudes were limited well below the basic insulation level (BIL) of the transformer in all cases where surge arresters were installed. Without protection by surge arresters, overvoltage amplitudes well exceeding the BIL have been observed at transformer windings when opening a nearby vacuum interrupter.

Mehdi Nafar, Ghahraman Solookinejad and Masoud Jabbari (2014) Compared IEEE and Pinceti Models of Surge Arresters. The main causes over voltages in power system are switching and lightning. The over voltages can damage the insolation of lines and equipment's connected to the power system. To protect of insulations and equipment of the power systems from the damaging effects by lightning over voltages, metal oxide surge arresters have been used. Because of dynamic behaviour of the surge arresters, they can't be simulated using non-linear resistors. Therefore, several models are proposed for simulate the dynamic properties of surge arresters. IEEE and pinceti models are the main models that proposed in order to simulate the dynamic behaviour of surge arresters. In this paper, for identification of surge arrester parameters, a novel algorithm has been proposed and then a comparison among IEEE model and pinceti model has been investigated. In this paper, IEEE and pinceti models of surge arresters are studied and compared. These models have been simulated using EMTP software. The mentioned models can estimated using a new method that proposed by authors. The proposed method is a common method and using this algorithm parameters of all models can determine. Both models can reproduce the property of metal oxide arresters precisely. In pinceti model, nine parameters must be estimated and in IEEE model, eleven parameters must be estimated, so pinceti's convergence speed is faster.

*****

CHAPTER 3
SYSTEM STUDY AND SIMULATION

3.1 Existing System and Associated Problem

The device which is used for the protection of the equipment at the substations against travelling waves, such type of device is called lightning arrester or surge diverter. In other words, lightning arrester diverts the abnormal high voltage to the ground without affecting the continuity of supply. It is connected between the line and earth, i.e., in parallel with the equipment to be protected at the substation. The substation experiences various overvoltages due to switching transient, resonance or electrostatic discharge. The failure of lightning arrestors due to overvoltage is serious problem. This paper suggest a method of using power diode across lightning arrestor to avoid failure . The MATLAB simulation model is prepared and conclusion has been drawn through waveform and data analysis.

When a travelling wave reaches the arrestor, its sparks over at a certain prefixed voltage as shown in the figure below. The arrestor provides a conducting path to the waves of relatively low impedance between the line and the ground. The surge impedance of the line restricts the amplitude of current flowing to ground. The lightning arrester provides a path of low impedance only when the travelling surge reaches the surge diverter, neither before it nor after it. The insulation of the equipment can be protected if the shape of the voltage and current at the diverter terminal is similar to the shape shown below in fig 1 and 2.

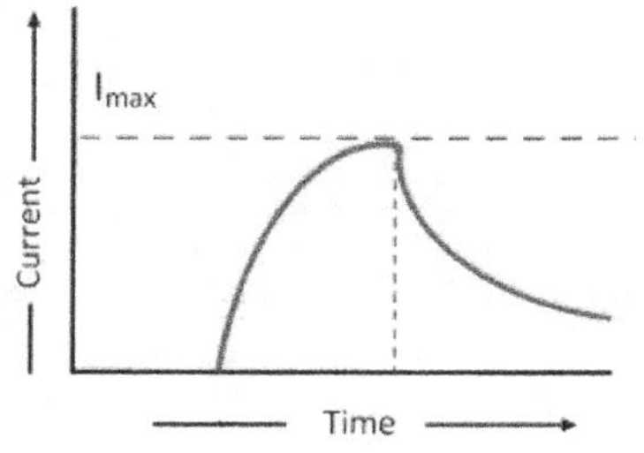

Current Characteristic

Figure 1

(Fig 1 and 2 voltage and current characteristics of L.A.)

An ideal lightning arrester should have the following characteristics;

1. It should not draw any current during normal operating condition, i.e., it sparks-over voltage must be above the normal or abnormal power frequency that may occur in the system.
2. Any transient abnormal voltage above the breakdown value must cause it to break down as quickly as possible so that it may provide a conducting path to ground.
3. When the breakdown has taken place, it should be capable of carrying the resulting discharge current without getting damaged itself and without the voltage across it exceeding the breakdown value.
4. The power frequency current following the breakdown must be interrupted as soon as the transient voltage has fallen below the breakdown value.

There are many types of lightning arrester which are used to protect the power system. The choices of the lightning arrester depend on the factor like, voltage and frequency of the line, cost, weather condition and reliability.

Location of Lightning Arrester

The lightning arrester is located close to the equipment that is to be protected. They are usually connected between phase and ground in an AC system and pole and ground in case of the DC system. In an AC system, separate arrester is provided for each phase.

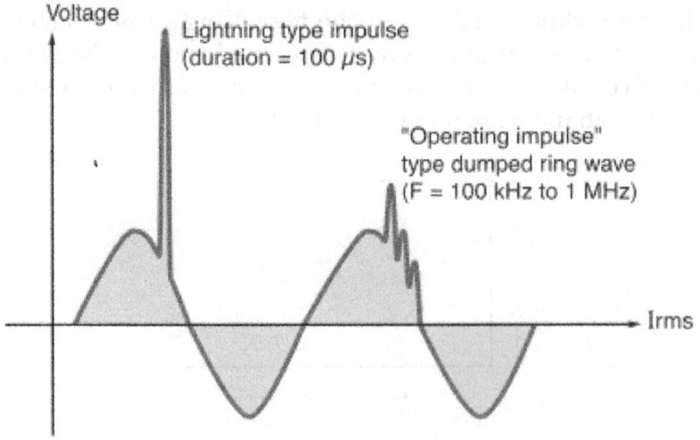

Figure 3: Examples of overvoltage

In an extra-high voltage AC system the surge diverter is used to protect the generators, transformers, bus bars, lines, circuit breakers, etc. In HVDC system the arrester is used to protect the buses, valves converter units reactors, filter, etc.

Overvoltage: overvoltage (in a system) is defined as any voltage between one phase conductor and earth or between phase conductors having a peak value exceeding the corresponding peak of the highest voltage for equipment.

An overvoltage is a voltage pulse or wave which is superimposed on the rated voltage of the network.

This type of overvoltage is characterized by

(see Fig. 4):

the rise time tf (in μs);

the gradient S (in kV/μs).

An overvoltage disturbs equipment and produces electromagnetic radiation. Moreover, the duration of the overvoltage (T) causes an energy peak in the electric circuits which could destroy equipment.

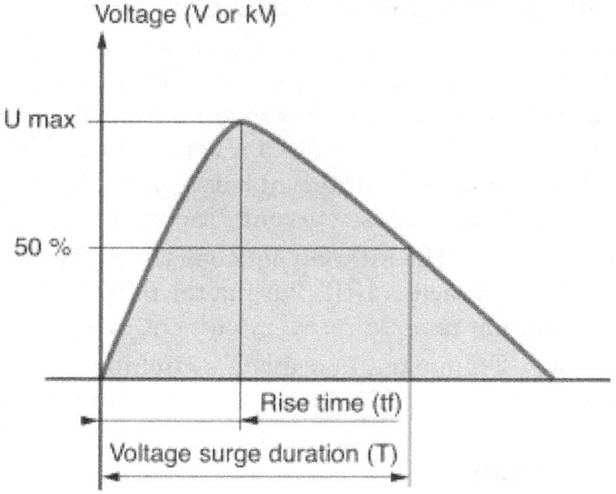

Figure 4: Main characteristics of an overvoltage

Four types of overvoltage can disturb electrical installations and loads:

1. **Switching surges:** high-frequency overvoltages or burst disturbance (see Fig. J1) caused by a change in the steady state in an electrical network (during operation of switchgear).
2. **Power-frequency overvoltages:** overvoltages of the same frequency as the network (50, 60 or 400 Hz) caused by a permanent

change of state in the network (following a fault: insulation fault, breakdown of neutral conductor, etc.).

3. **Overvoltages caused by electrostatic discharge:** very short overvoltages (a few nanoseconds) of very high frequency caused by the discharge of accumulated electric charges (for example, a person walking on a carpet with insulating soles is electrically charged with a voltage of several kilovolts).

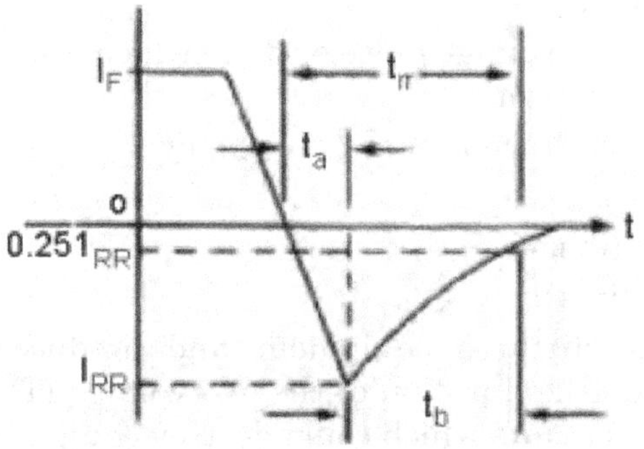

4. **Overvoltages of atmospheric origin**
   - Power Diode Characteristics: The reverse recovery characteristics of the Power diode are shown in the figure 5. From the figure, we can understand the turn off characteristic of the diode. The Reverse recovery time Trr is the (fig. 5 reverse recovery characteristics of the Power diode)time interval between the application of reverse voltage and the reverse current dropped to 0.25 of IRR.
   - Parameter ta is the interval between the zero crossing of the diode current to it reaches IRR. Parameter tb is the time interval from the maximum reverse recovery current to 0:25 of IRR.
   - The lower Trr means fast diode switching. The ratio of the two parameters ta and tb is known as the softness factor SF.

3.2 Circuit Diagram

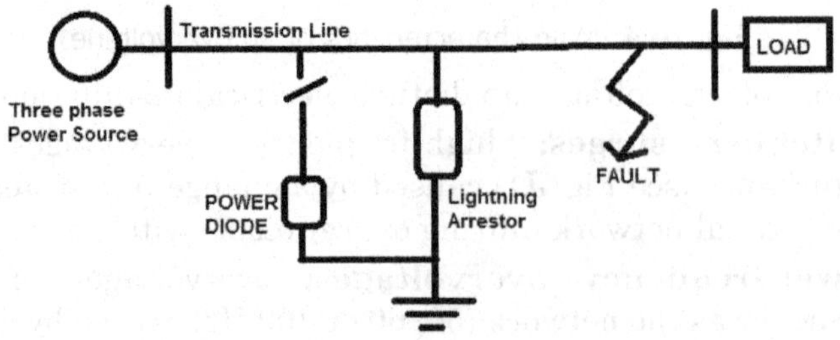

Datasheet Parameters:

For power diodes, a data sheet will give two voltage ratings. One is the repetitive peak inverse voltage (VRRM) and the other is the non-repetitive peak inverse voltage.

The non repetitive voltage (VRM) is the diode's capability to block a reverse voltage that may occur occasionally due to over voltage surge.

The data sheet of a diode normally specifies three different current ratings. They are: (1) Average current; (2) RMS current; and (3) Peak current. A design engineer must ensure that each of these values are never exceeded.

3.3. MATLAB Simulation

The 132 KV three phase transmission systems is simulated in MATLAB using Simulink. The model consist of 132 kv generation system, transmission line pi-section of 100 km length, the metal oxide varistor (MOV) type lightning arrestors are connected between phase to ground with leakage current measuring device.

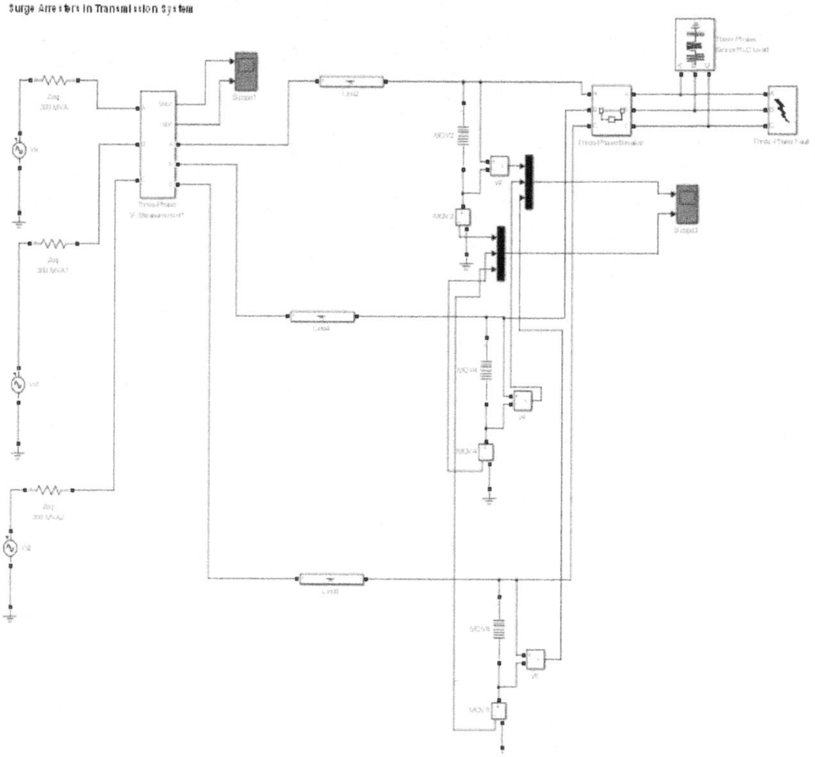

The transmission systems emploes three phase R-L Load and connected through circuit breaker. On load side a three phase fault is simulated. Supply system voltage, current is measured. The voltage across L.A. and leakage current is also measured using scopes. The two anti-parallel power diodes on each phase across L.A. is also connected. Practical MPSEB Data has been used to obtain simulation results

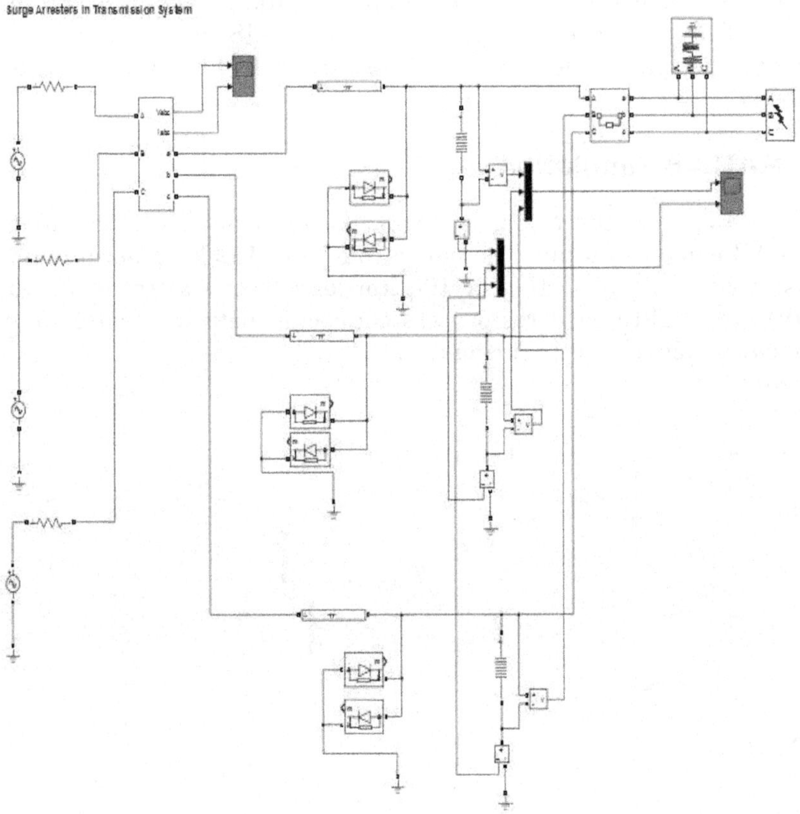

*****

CHAPTER 4
RESULTS AND DISCUSSION

WAVEFORMS

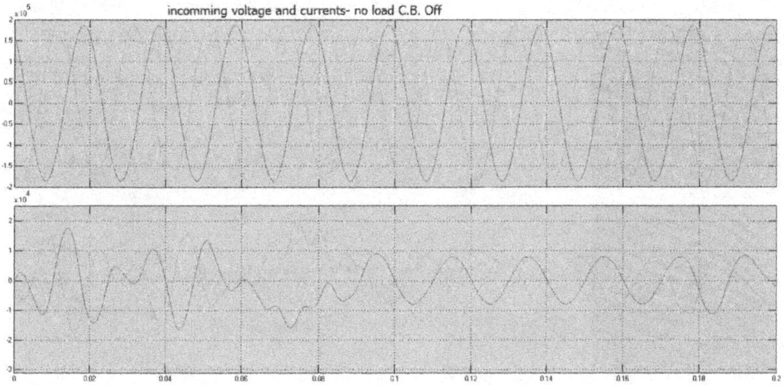

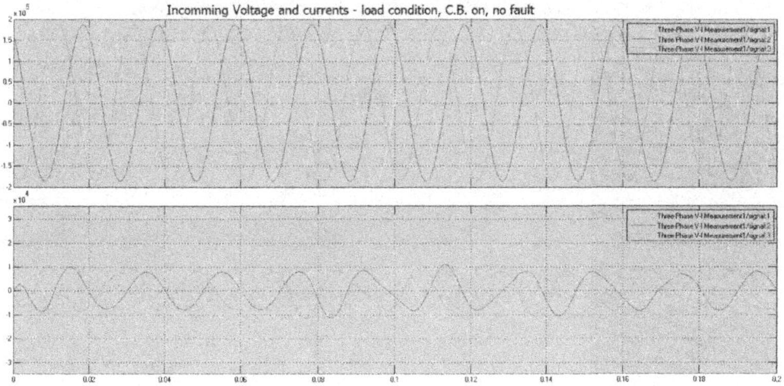

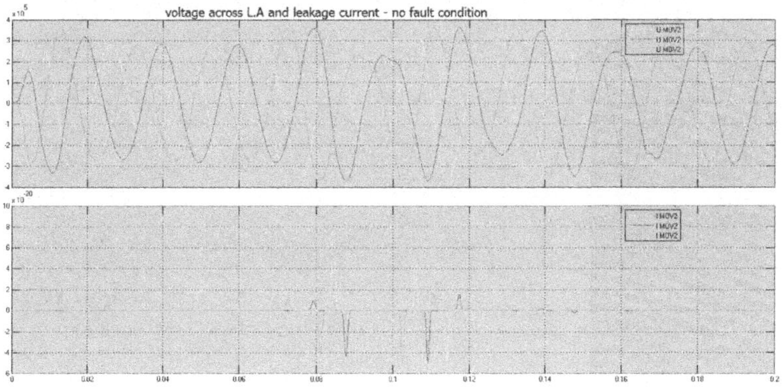

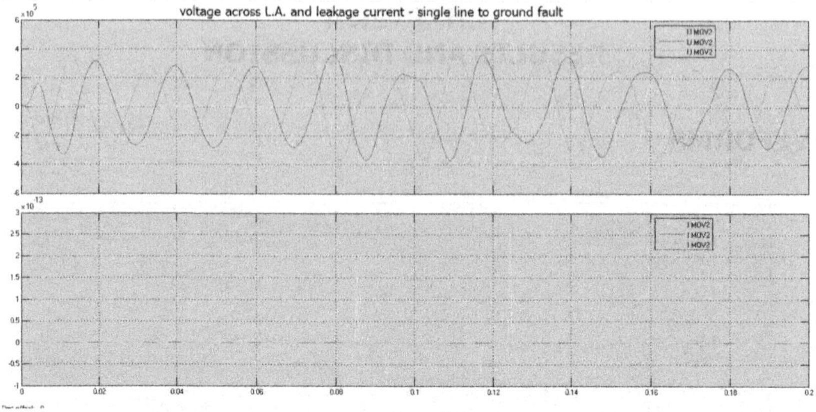
voltage across L.A. and leakage current - single line to ground fault
U MOV2
U MOV2
U MOV2
I MOV2
I MOV2
I MOV2

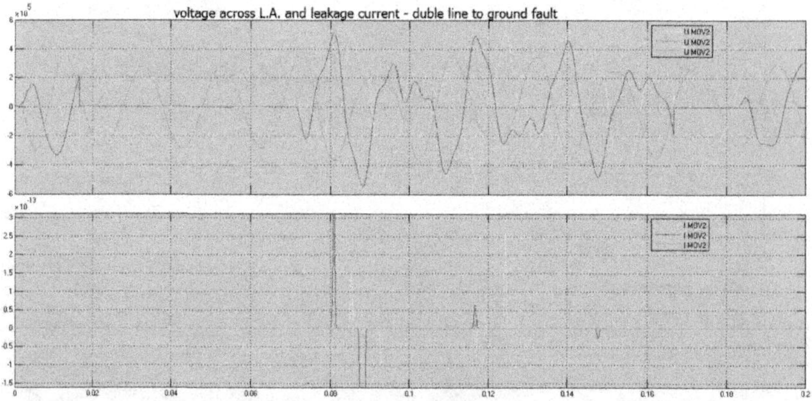
voltage across L.A. and leakage current - duble line to ground fault
U MOV2
U MOV2
U MOV2
I MOV2
I MOV2
I MOV2

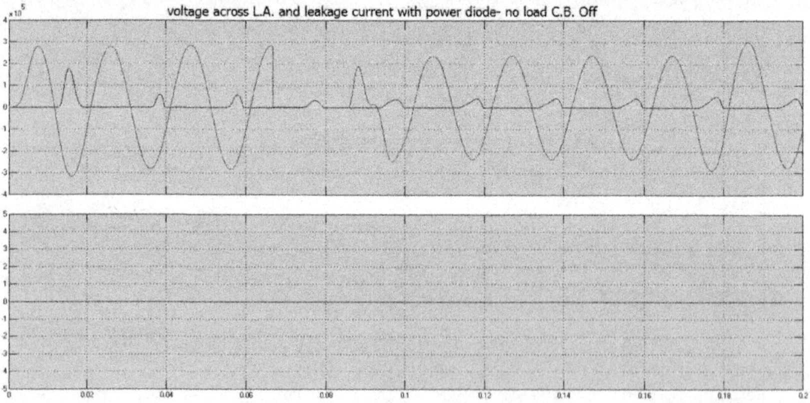
voltage across L.A. and leakage current with power diode- no load C.B. Off

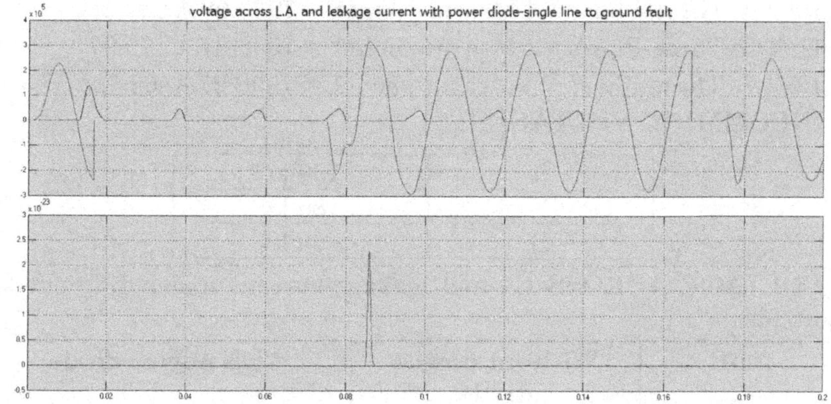

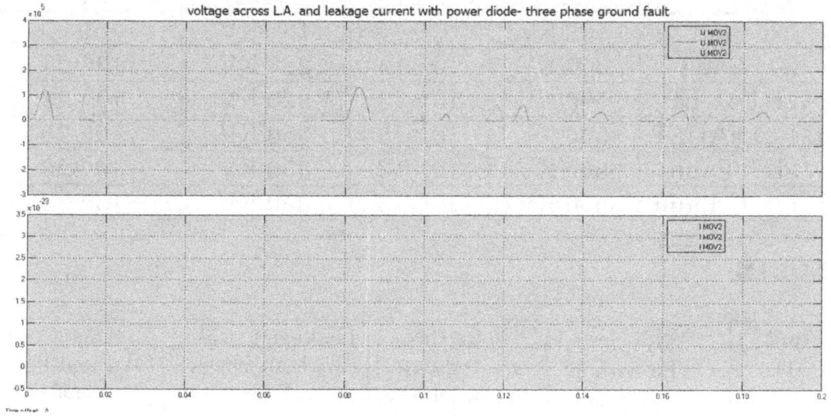

CALCULATED VALUES

STEADY STATE PARAMETER MEASUREMENTS:

1: 'U_v6 ' = 283226.36 V -142.04°

2: 'U_v2 ' = 283699.81 V -22.44°

3: 'U_v4 ' = 283226.36 V 97.96°

4: 'U A: Three-Phase V-I Measurement1' = 185995.51 V -0.00°

5: 'U B: Three-Phase V-I Measurement1' = 185995.52 V 120.00°

6: 'U C: Three-Phase V-I Measurement1' = 185995.52 V -120.00°

7: 'I_IMOV2 ' = 0.00 A 0.00°

8: 'I_IMOV4 ' = 0.00 A 0.00°

9: 'I_IMOV6 ' = 0.00 A 0.00°

10: 'I A: Three-Phase V-I Measurement1' = 7937.08 A 55.52°
11: 'I B: Three-Phase V-I Measurement1' = 7996.27 A 175.97°
12: 'I C: Three-Phase V-I Measurement1' = 7996.27 A -64.03°

INCOMMING VOLTAGES

R	186 kv
Y	186 kv
B	186 kv

VOLTAGE ACROSS LA and leakage current under different fault conditions

fault	Without power diode		With power diode	
	V	I	V	I
No fault	187 KV	0	187 KV	0
L-G Fault	350 KV	10^-13	228 KV	10^-23
L-L FAULT	500 KV	10^-13	320 KV	10^-23
L-L-G FAULT	540 KV	10^-13	350 KV	10^-23
L-L-L Fault	550 KV	10^-13	350 KV	10^-23
L-L-L-G Fault	550 KV	10^-13	350 KV	10^-23

RESULTS

Fault Type	Volta Ges On La without Power Diodes In Kv	Volta Ges On LA with Power Diodes In Kv	Leakage Current without Power Diodes	Leakage Current Without Power Diodes
No fault	76.21	76.21	1.1 mA	1.1 mA
R Phase to earth	210.30	100.72	45 mA	2.6 mA
R-Y Phase	220.36	110.72	42 mA	2.1 mA
R-Y-Earth	214.44	104.17	48 mA	2.9 mA
R-Y- B Phase	244.10	101.88	43 mA	2.1 mA
R-Y-B-earth	275.30	145.72	52 mA	2.8 mA

Conclusion

It has been observed that using power diodes combination across lightning arrestor the overvoltages are reduced to much extents also the leakage currents also decreases. Thus it reduces the failure of lightning arrestor and increasing the life span. It is also observed that using power diodes with suitable resistance values also reduces overvoltages during different fault conditions. Modern metal oxide arresters provide markedly superior protective characteristics and energy absorption capability, compared to previous generation arresters. The application and selection of metal oxide arresters requires a thorough review of the power system, including voltage, system stresses, switching surges, grounding method and MCOV.

Bibiliography

[1] Ministry of Railway, govt of india – Condition Monitoring of Lightning Arrestor, 2016

[2] C.A. Christodoulou M.V. Avgerinos- Measurement of the resistive leakage current in surge arresters under artificial rain test and impulse voltage subjection,IET Science, Measurement and Technology, 2009

[3] Volker Hinrichsen -Monitoring of High Voltage Metal Oxide Surge Arresters,VI Jornadas Internacionales de Aislamiento Eléctrico,1997

[4] M. Gumede and G. Frederick d'Almaine- Surge Arrester Faults and Their Causes at EThekwini Electricity,International Journal of Electrical Energy, Vol. 2, No. 1, March 2014

[5] M Darveniza, TK Saha and S Wright-Studies of in-service and laboratory failures of metaloxide distribution surge arresters,Journal of Electrical & Electronics Engineering, Australia, 2003

[6] https://www.nist.gov/sites/default/files/documents/pml/div684/All_papersPart8.pdf

[7] The Application and Selection of Lightning Arresters By Larry Pryor Fernando Gontijo Bernardes et all - Automatic detection of fault patterns in lightning arresters, 2017 International Symposium on Lightning Protection (XIV SIPDA), Natal, Brazil, Liu, gao, meng, tao - design and monitoring of online device for arrestor in substation, IEEE 2017

[8] L. Liljestrand and E. Lindell- Efficiency of surge arresters as protective devices against circuit breaker induced overvoltages, IEEE 2015

[9] Doan Duc Tung- Over-Voltage Prevention Solutions for 220 kV Distribution Substation to Enhance Power System Stability: A Case Study in Vietnamese Power System, International Journal of Applied Engineering Research ISSN 0973-4562 Volume 12, Number 18 (2017)

[10] Nikolay Nikolov, Neli Dimitrova, Anton Georgiev and Margreta Vasileva-Reliability Assessment of Electricity Distribution Substation Surge Protection System, 40th International Spring Seminar on Electronics Technology, IEEE 2017

[11] Valdemir S. Brito, George R. S. Lira, Edson G. Costa, Marcelo J. A. Maia-A Wide-Range Model for Metal-Oxide Surge Arrester, IEEE 2016

[12] 14.Leela AM, Ravi KN, Muralidhara, VAsudeva, Shivkumara-Simulation and Analysis of ZnO block

[13] Characteristics Using Matlab, 2017 IEEE International Conference on Intelligent Techniques in Control, Optimization and Signal Processing

[14] Mehdi Nafar, Ghahraman Solookinejad and Masoud Jabbari-Comparison of IEEE and Pinceti Models of Surge Arresters, *Research Journal of Engineering Sciences 2014*

[15] M. Khodsuz and M. Mirzaie-Condition Assessment of Metal Oxide Surge Arrester Based on Multi-Layer SVM Classifier, Iranian Journal of Electrical & Electronic Engineering, Vol. 11, No. 4, Dec. 2015

[16] Nurlan Ibrayev, Aigerim Uyzbayeva — Application of nonlinear surge arrestors for lightning protection of overhead power lines and Modelling of overvoltage in dead-end substation, 4th International Conference on Power Engineering, Energy and Electrical Drives Istanbul, Turkey, 13-17 May 2013.

*****

SECTION D
ANALYSIS OF POWER QUALITY DISTURBANCE DUE TO VOLTAGE SAG ON 33/11 KV NETWORK WITH DIFFERENT TRANSFORMER CONNECTIONS

Abstract

In Present era with the continuous increase in consumption of power supply, people are also concern about power quality. Both utilities as well as end users are very much aware of power quality. The term power quality in itself cover all types of power quality disturbances. Present generation aspire for such power plant design which can meet their requirements as well as can fulfil future needs also. To design such power plant, power engineers have to devote their time and work towards improving the quality of power without sacrificing consumers demand. In today's life style, consumers use so many electronic non-linear devices for their comfort. These devices are one of the main reason of power quality disturbances. In this dissertation work power quality disturbances due to voltage sag has been discussed.

Voltage sag is one of the most important point for disturbing power quality of the system. Voltage sag occur due to faults such as SLG, DL, DLG and 3-Φ on 11 kv line and how transformer connections effect these generated voltage sag is taken as the main concern. To analyze this software used is MATLAB/SIMULINK.

Keywords: Power quality disturbances, Voltage sag, Voltage Sag Indices, Area of Vulnerability, Rectangular and Non-Rectangular sag.

CHAPTER 1
INTRODUCTION

1.1 Introduction

The term power quality is not so old term. It comes to highlight in recent decades since in late 1980s. Nowadays people are becoming so much aware of this term power quality. Consumers need continuous power supply but not at the cost of quality. They demand good quality continuous power supply. Power quality in itself covers each and every type of power quality disturbances. Utilities and end users both wants good quality of power. So power engineers have to work not only for fulfilling increase in supply demand but also have to design such system which can meet all demands without sacrificing quality of power.

The major causes of increased power quality concerned are -

1. Present generation load equipments, with microprocessor-based controllers and non-linear devices are very sensitive to power quality.
2. To obtain good efficiency continuous use of high efficiency, adjustable motor speed drives, capacitors for power factor correction in order to reduce losses. This in turn give birth to power quality disturbances in the system.
3. Sags, interruptions, transients etc are the big challenges for the utilities. For which utilities have to be more concern about the quality of power.
4. In interconnected network system, failure of any component or element became the important consequences of power quality.

All these above mentioned points increase the importance of study of power quality disturbances. Here, in this dissertation work power quality disturbances due to voltage sag has been discussed. Voltage sag is one of the main type of power quality disturbances. Here, in this analysis work all parts of voltage sag such as definition, effects, causes, characterization all these things are discussed. Apart from all these basic points of discussion one more thing is highlighted that is area of vulnerability, equipment's ride through capability against voltage sag considering both rectangular and non-rectangular sag.

This work includes discussion of voltage sag indices in order to calculate voltage sag by using RMS value of voltage. In this thesis work 33/11 kv network of Madhya Pradesh Power Transmission company Ltd. feeding Shaktinagar is taken as case study. First the line is modeled by using MATLAB/SIMULINK and different types of faults are applied. By applying different types of faults on various transformer connections, system behaviour has been studied out.

1.2 Objective of Thesis

The main objective of this thesis work is as follows.

1. To analyze voltage sag on 11 kv line.
2. To study area of vulnerability and equipments ride through capability.
3. Various transformer connections behaviour on different type of fault.

1.3 Overview of Thesis

The complete overview of thesis is as follows.

- Chapter 1 deals with brief introduction of the dissertation work.

- Chapter 2 deals with brief introduction of power quality disturbances and voltage sag, its causes, effects and characterization.
- Chapter 3 deals with area effected by sag, equipments ride through capability, overview of IEEE.
- Chapter 4 deals with RMS calculation method, how to obtain numerical value of sag using voltage sag indices.
- Chapter 5 deals with different types of faults causing voltage sag.
- Chapter 6 deals with comparison of different type of fault on various transformer connections.
- Chapter 7 deals with conclusion and future scope.

1.4 Literature Survey

This dissertation work would be possible without the help of these research papers, websites and reference books. To make it achieve to a successful end these literature survey helps a lot.

(1) J. Caicedo, Felipe Navarro, Edwin Rivas and Francisco Santamaria "Voltage sag characterization with MATLAB/ SIMULINK." in IEEE 2012.

In this paper a method for voltage sag characterization by using MATLAB/ SIMULINK has been suggested. In this fault simulation, VS magnitude calculation is done by using sim power system tools. In this over-current protection has been simulated and also model developed in MATLAB/ SIMULINK. Phase angle jump in case of induction motor has been discussed with the simulation model.

(2) Suresh Kamble, Chandrashekhar Thorat "Voltage sag characterization in a distribution systems: A case study" April, 2014 Journal of power and energy engineering.

Here, In this paper VS due to SLG, DL and DLG faults are characterize by using symmetrical component analysis and their effect on magnitude and phase angle jumps on each phase being analyzed. In this VS has been calculated by RMS method for which a model has been simulated on MATLAB/ SIMULINK and different types of faults applied on it and their VS behaviour or effects have been studied out. These effects are studied on the basis of magnitude, duration and phase angle jump.

(3) D.J.Won, Seon-Ju Ahn and Y.Chung "A new defination of voltage sag duration considering the voltage tolerance curve. (2003) IEEE Bologna Power Tech Conference, Bologna, Italy.

In this research paper a new VS duration which consider voltage tolerance characteristics of each electrical devices has been discussed. In conventional method, VS for rectangular sag has been consider whenever the sag is non-rectangular, VS has always been overestimated. To overcome from this, method proposed takes into account minimum voltage of voltage tolerance curve of each equipment. This new method have been applied on non-rectangular sag generated from induction motor starting and compared with the maximum time to calculate VS effect on equipment.

(4) E.Styvaktakis, M.J.H.Bollen "Automatic classification of power system events using RMS voltage measurements" (2002) IEEE Power Engineering Society Summer Meeting.

In this paper PQ monitoring in depth has been discussed. For this a method have been suggested for automatic power system event classification using RMS voltage measurement and results show that this classification can be applied on specific events and its limitations have been discussed. In this events covered are energizing, non-fault interruption, fault interruption, transformer saturation, induction motor starting, step change and faults.

(5) P.Heine, M.Lehtonen "Voltage sag distributions caused by power system faults" IEEE Transaction on Power System.

In this paper VS distribution on different voltage level is taken into consideration. Share of different types of faults at each voltage level and sag propagation throughout the power system has been studied. By this it is concluded that origin of sag in urban and rural areas are different. This paper helps during planning of power system to mitigate sag from different parts of power system.

(6) S.Khokhar, A.A Mohd Zin, A.S Mokhtar, NAM Ismail " MATLAB/ SIMULINK based Modelling and simulation of power quality disturbances" IEEE (2014)

In this paper, simulation and Modelling of PQD due to exploitation of various load types has been discussed. PQD created by using parametric equations as well as power system models in MATLAB/ SIMULINK. PQD of voltage magnitude variation such as sag, swell and interruption are created by different types of faults and heavy loads. PQD of frequency variation type such as harmonics are generated by using converter. PQ waveforms obtain by this are useful in checking the performance of new automatic classification algorithm.

(7) IEEE Recommended practice for monitoring electric power quality 1159-2009.

In this, standards of IEEE on monitoring of electrical PQ has been discussed. This is proved to be very helpful in understanding PQD due to VS and equipments ride-through capability compatible with VS.

*****

CHAPTER 2
INTRODUCTION TO POWER QUALITY DISTURBANCE AND VOLTAGE SAG

2.1 What is Power Quality Disturbance?

Day by day people are adopting higher living standard values. For their own comfort, they are using so many electronic devices. Although these devices have made our life easy and comfortable but at the cost of power quality disturbance (PQD). Many of these electronic devices are very sensitive to the power which is being supplied to them. Quality of power is directly linked with the voltage in the power system, in turn stability of power quality is referred to voltage stability in the power system [17]. Power quality disturbances arise due to harmonics, voltage regulation, interruptions etc. Among all these, the reason which is going to be discussed here is voltage sag. PQ of the system gets disturbed due to two main factors, first is electronic device and second is fault in the system [11]. Those electronic devices which are not linear in nature which require huge current to start, gives birth to PQD. And on the other hand, faults occur in the system either from utility side or from end user side or some natural reasons also degrade the quality of power. Little bit deviation of current or voltage from the specified value is referred to as PQD.

In present era, power engineers need to plan the power system design in such a way that the system can sustain its PQ even after facing all the above said problems. Along with that the system should also be compatible with the increasing demand of power supply. Quality of power depends on the system design, reliability and efficiency so study of PQD is important. PQD affects industrial as well as large commercial customers. To understand the system behaviour, PQD plays a vital role or referred to as main basis of concern, to give information about system behaviour. Two of the important divisions of PQD are variations and events. Variations need to be monitored as they are steady-state disturbances whereas events are sudden disturbances in the system.

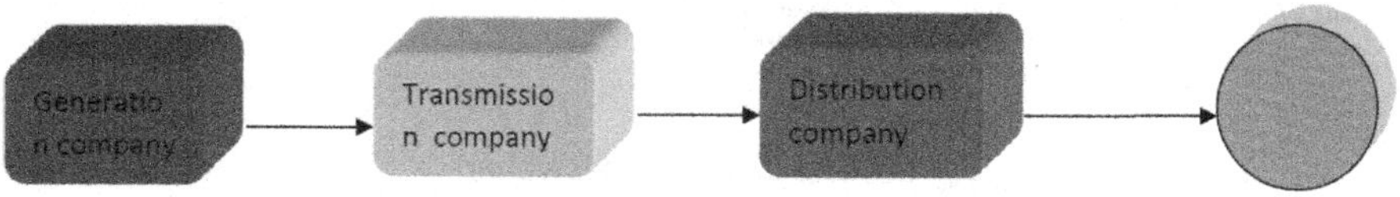

Figure 2.1: Power system structure

So, to maintain the power quality, study or analysis of PQD is very important. Analysis of PQD includes all types of disturbances but here we are going to discuss about voltage sag only. PQD due to VS on 11 kv line is analyzed on different transformer connections by applying different types of fault with the help of MATLAB/ SIMULINK.

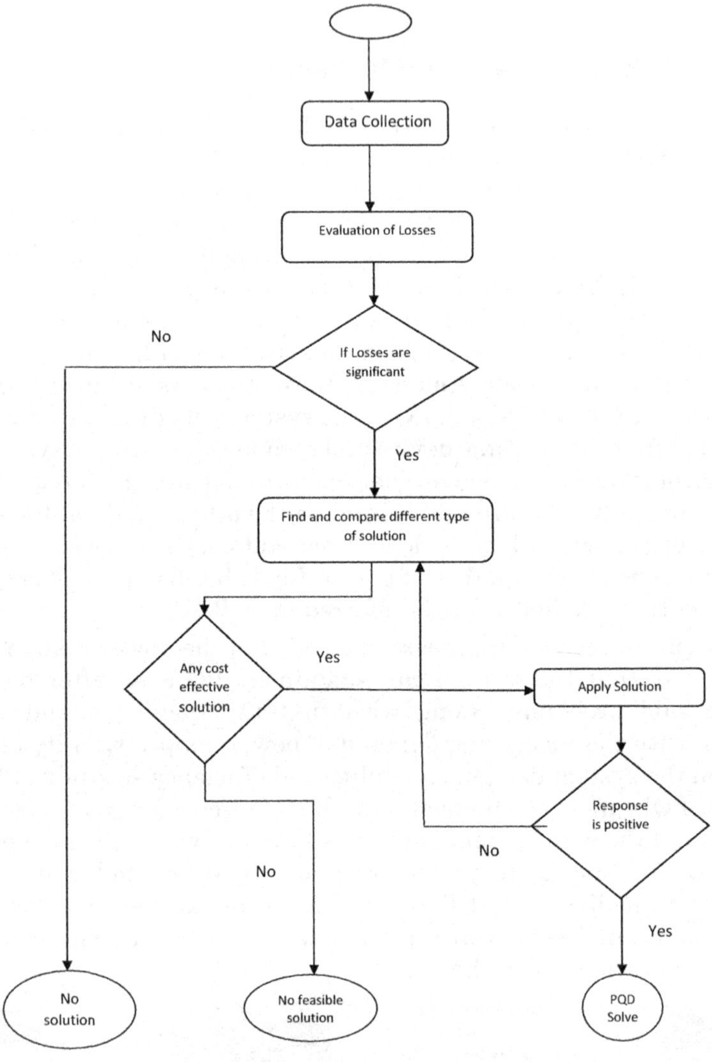

Figure 2.2: PQD Flowchart

2.2 What is Voltage Sag?

The term voltage sag as the name suggests, is drop in voltage. VS distorts the voltage profile of the system. Voltage sag is also referred as voltage dip. Voltage dip is the term given by IEC while the term "voltage sag" is used by North American power quality commission [21]. VS creates disturbance in the pure sinusoidal waveform of voltage which in turn deteriorates the power quality of the system. VS occurs for very short duration of 0.5 to 30 cycles. It is decrease in RMS voltage of about 90 % to 10 % of nominal voltage [4][3][1]. Although VS occurs for very short duration but its effects deteriorates system with time decreases life period of the system. Probability of large VS is less than small VS due to electrical supply configuration [9]. VS is drop in voltage which is being supplied. VS badly affects the voltage stability of the system. To calculate out VS, drop in RMS value of voltage is the main concern. By calculating out RMS voltage drop, percentage VS can be determined.

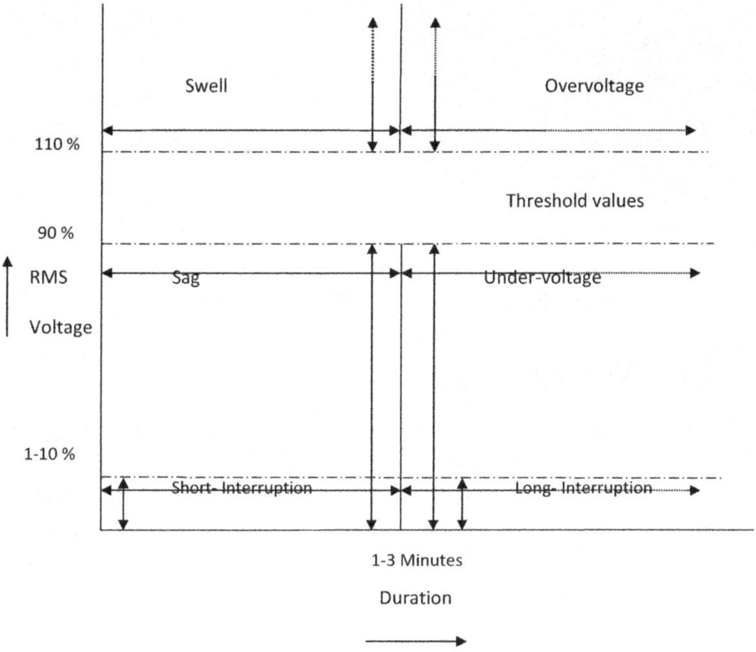

Figure 2.3: Types Of Voltage Variation Duration Curve

Here, we study the voltage sag relation with different transformer connections. By applying different types of faults, sag in voltage is developed in the system. Voltage sag analysis gives us complete details of system behaviour when voltage drops from the specified value or threshold value. VS affects both domestic as well as industrial customers but comparatively industrial customers have to face more problems due to VS.

Table 2.1: VS Classification

Type of Sag	Duration	Magnitude
Instantaneous	0.5-30 cycles	0.1-0.9 pu
Momentary	30 cycles-3 sec	0.1-0.9 pu
Temporary	3 sec-1 min	0.1-0.9 pu

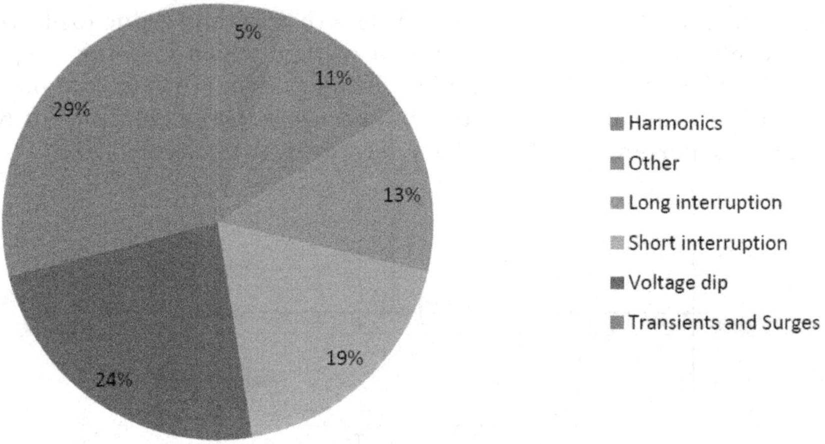

Figure 2.4: PI-Chart for different types of PQD

2.3 Causes of Voltage Sag

VS occurs for short duration of time due to short circuit in the system. Causes of VS are from both sides that is utility side as well as end users side. VS arises due to utility side because of continuous switching on and off of equipments connected in the system with the help of circuit breakers, disconnection of reclosures and switches [5][10]. VS occurred due to end users side is because of use of non-linear loads starting of large motor which requires heavy current, disturbs the sinusoidal waveform of voltage.[11][19] Apart from these, there are some natural reasons which causes voltage drop such as lightning strikes, car hitting power pole, animal and trees touching or falling on power lines etc [18]. All these faults are categorized under single line to ground or double line, three phase to ground or three phase faults. Variation in voltage is the result of significantly varying loads connected in the system. Mostly it is observed that VS occurs due to SLG fault in the system more as compared to other types of fault. Some of the causes VS are listed below -

1. Electric motors need more current at the starting time than that at the running time, which causes VS in the system.

2. During SLG fault, until the protective switch gear comes to its action there will be VS in the line.
3. Accidents in power line such as lightning or falling of an object.
4. Excessive load or sudden load change can cause VS.
5. Depending on transformer connections, transformer energizing also gives birth to VS.
6. VS is seen when air-conditioner, refrigerator or furnace fan start.

Power system design must be that much robust that it can bear all above mentioned causes of VS without any quality disturbances in the system. Power engineers need to work on this, to make it possible practically.

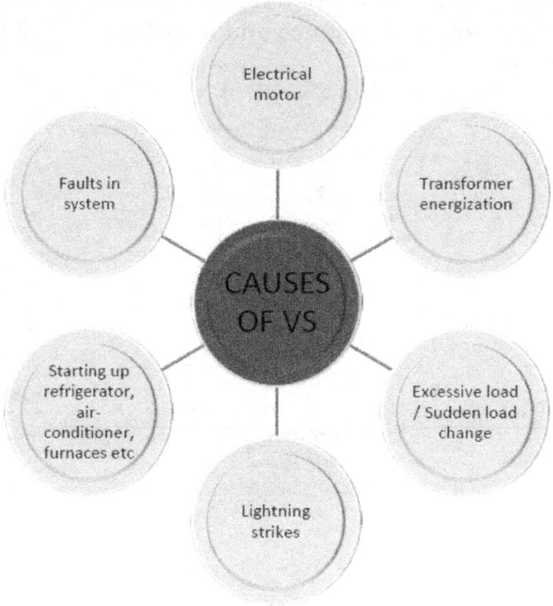

Figure 2.5: Causes Of VS

2.4 Effects of Voltage Sag

VS experienced by low voltage consumer is more as compared to other consumers even though sag is distributed in the whole system [4]. This happens only because of network characteristic features, sag in some part is high and at other part, it is low or many a times negligible. VS results in breakdown of plant, increase in maintenance cost, production loss etc. It badly affects sensitive equipments such as adjustable speed drives, diagnostics systems, computer controller equipments etc. Effects of VS seen are as follows [12]-

1. System losses increase.
2. Reactive power requirement increase.
3. Poor power quality.
4. Equipment malfunction and decrease in their life time.
5. System becomes uneconomical.

The economical consequences which an industry has to face due to VS can be calculated by summing up the cost of repairment and the cost of production loss [13]. VS is always confused with the term interruption, but there are remarkable differences between the two. Interruptions occur for 1 minute, decrease in RMS voltage to less than 10 % of nominal value. Although VS does not cause severe damage at the time when it occurs but deteriorates the system performance and also decrease its life time.

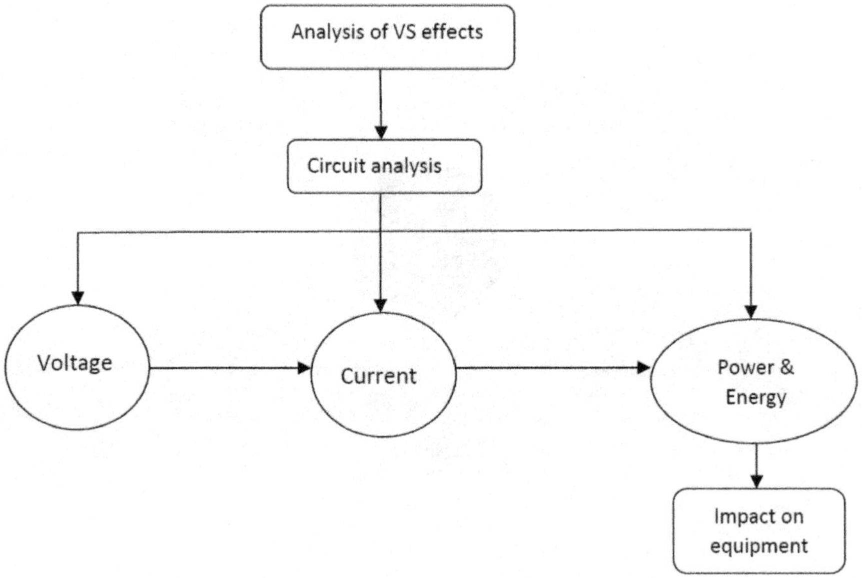

Figure 2.6: Flowchart for analysis of VS effects

2.5 Characterization of VS

To analyze PQD study of VS characterization is important. Basically VS is characterized on the basis of magnitude and duration, one more term used to characterize is phase angle jump. Apart from all these three, VS is also characterized by unbalance of sag [8][20]. So, VS is characterized in four classes that is magnitude, duration, phase angle jump and last but not the least unbalance of sag. All these are described in brief as follows -

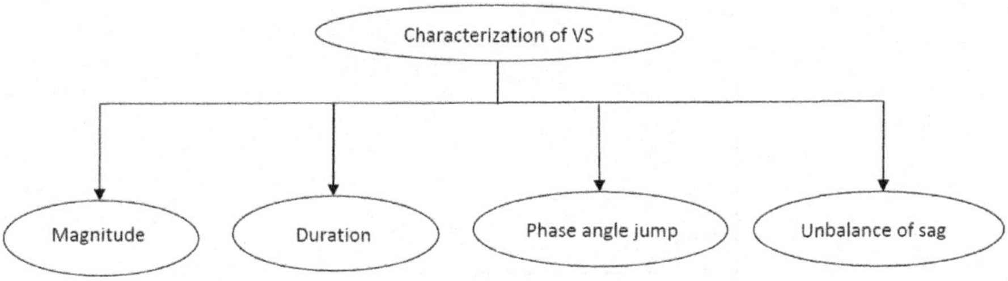

Figure 2.7: Characterization chart

2.5.1 Magnitude

The term magnitude is defined as the last remaining voltage during the sag event. To determine magnitude of sag, RMS voltage is calculated and also by calculating peak voltage. In three phase system, if fault is symmetrical in nature than the lowest voltage remained in the system is consider and when the fault is unsymmetrical in nature then the lowest voltage among one of the three phases are considered to calculate magnitude of sag [2]. Magnitude of sag depends on system configuration, protective devices or fault clearing devices and fault location. Magnitude is calculated by formula

$$V_{sag} = [\, F_Z/(S_Z+F_Z) \,]\, E \quad (2.1)$$

Where,

V_{sag} = VS magnitude

F_Z = Fault impedance

S_Z = Source impedance

Take pre-fault voltage as 1 pu. Therefore, E=1

$$V_{sag} = [\, F_Z/(S_Z+F_Z) \,] \quad (2.2)$$

2.5.2 Duration

Duration is defined as the time period for which VS event occur that is RMS value of voltage drop below the permissible value. VS duration completely depends on action taken by protective devices to clear the fault from the system [9]. In distribution system, faults are cleared slowly as compared to transmission system. So depending on fault location, duration of VS in distribution system is more than that of transmission system. VS duration starts, when one of the three phase RMS voltage values goes down to the threshold value and ends with all the three phase RMS voltages are regained. So, to determine VS duration all the three phase RMS voltages are to be taken into consideration. Duration is further classified as momentary, instantaneous and temporary [15].

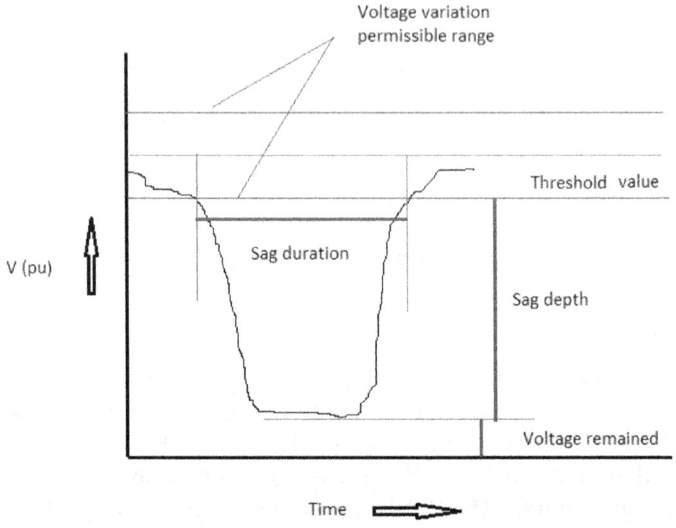

Figure 2.8: Sag duration curve

2.5.3. Phase Angle Jump

The phase angle before the VS occurred and the phase angle during the VS event, both of them together gives the phase angle jump details. Phase angle jump is located in time domain plot as a shift in voltage zero-crossing. In case of transmission system, no phase angle jump is being observed during faults even when source and feeder X/R ratio are same or equal but this does not happen in the case of distribution system. Phase angle analysis becomes more complex in case of unsymmetrical faults, because in such faults single phase load experiences quite significant amount of phase angle jump even for same X/R ratio.

For phase angle jump calculation formula used as:

$$\bar{V}_{sag} = [\ \bar{F}_Z / (\bar{S}_Z + \bar{F}_Z)\ ] \tag{2.3}$$

Where,

$$\bar{S}_Z = R_S + jX_S \tag{2.4}$$

$$\bar{F}_Z = R_S + jX_F \tag{2.5}$$

$$\Delta\Phi = \arg(\bar{V}_{sag}) \tag{2.6}$$

If

$$X_S / R_S = X_F / R_S \tag{2.7}$$

There is no phase angle jump in this case (1).

2.5.4 Unbalance of Sag

According to the type of fault sag is of two types symmetrical sag and unsymmetrical sag. Whenever three phase faults occurs sag effect in all the three phases will be experienced but when ever faults like SLG or DL occur sag effect in all the phases are not equal that is symmetrical [2]. This is called unbalance of sag, when the effect of sag is not experienced by all the three phases equally. Due to symmetrical faults, sag occurred will propagate through transformers without any change in transformer but when ever it occurred due to unsymmetrical fault it strongly affects the transformer connection.

*****

CHAPTER 3
STUDY OF AREA EFFECTED

3.1 Importance and Process of Calculating Effected Area

For power engineers to design the whole power system it is very important to have the data of area effected by sag, for this purpose study of area effected by sag is important. This study helps in decision making process by which engineers can estimate about the ratings of power system equipments. To establish compatibility in between supply system and the consumers following important steps must be taken.

1. Find out the characteristics and number of voltage sag events.
2. Find the equipments those which are sensitive to the sag event.
3. Search for best economical solution in order to improve the performance of system and also immunity of the devices installed in the system.

All these above mentioned steps help in making the proper co-relation between the system and consumers. The study of area effected by the sag is termed as area of vulnerability (AV). This study of area effected, at a small scale helps in minimizing its effect on the system. So, for making decision regarding ratings of the devices to be installed, this study is proved to be quite beneficial.

Table 3.1: VS Immunity Ranking Of Some Equipments

VS Immunity Ranking	Area	Overall Percentage
1	EMO circuit	47 %
2	DC power supply	19 %
3	Vaccum pumps	12 %
4	3-Φ power supply	12 %
5	Turbo pumps	7 %
6	AC drives	2 %

3.2 Area of Vulnerability (AV)

The term area of vulnerability is basically concerned with the minimum amount of voltage sag an equipment or installed devices can withstand, in simple words equipment's ride-through ability against voltage sag [6]. AV can also be called as voltage sag immunity or equipment's limit of susceptibility. This is very beneficial from the point of view to visualize the remaining voltage in the network due to short circuit fault in network. AV is taken as the zone of distribution network where short circuit causes VS, with a particular value or a short duration interruption. To determine the actual number of voltage sag combine the AV with the fault performance expected in that particular portion of circuit or say system. Expected fault performance is figured out by historical data. To calculate AV the total area of circuit which come under the effect of voltage sag which makes the equipment, drop in RMS voltage below the ride-through limit. To reduce the voltage sag effects we have to plan for reducing AV. Power engineers, with the help of AV can make so many decision to design the system robust and to ensure good efficiency from the system.

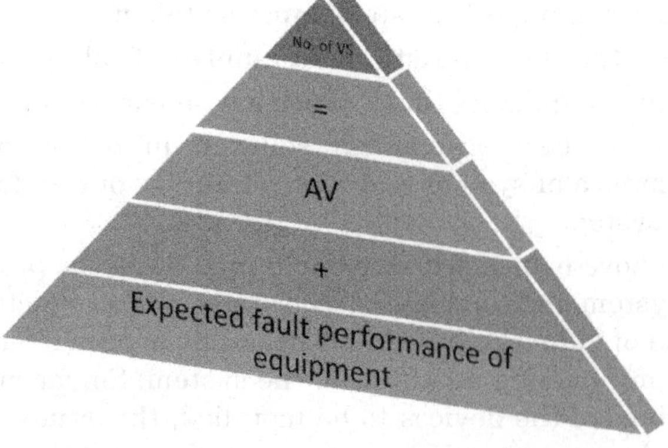

Figure 3.1: Determination of number of VS

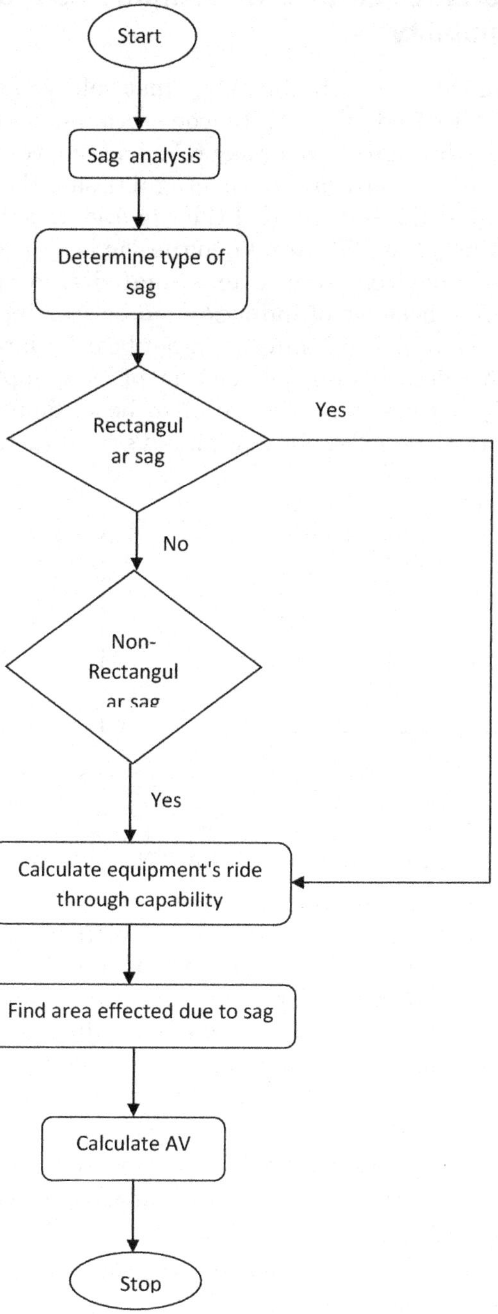

Figure 3.2: Flowchart for AV

3.3 Characteristic Features of Equipments Ride - Through Capability

Calculating ride-through ability against voltage sag is one of the most important part to find out AV. In conventional method for estimating voltage sag ride-through curve or say tolerance curve, only rectangular sag are taken into account. In case of rectangular sag, the tolerance curve has very sharp drop at the initiation of the sag and also it has very sharp rise at the ending time of sag [7]. But when the sag is non-rectangular, VS effect on equipments or devices can be overestimated. Generally non-rectangular sag are generated because of induction motor starting. In rectangular sag different type of equipment's effect can not be calculated out. Sensitivity of equipment to VS depends on specific type of load, applications and control settings. Many a time it is often find to be difficult to judge which VS characteristic is responsible for equipment's malfunction.

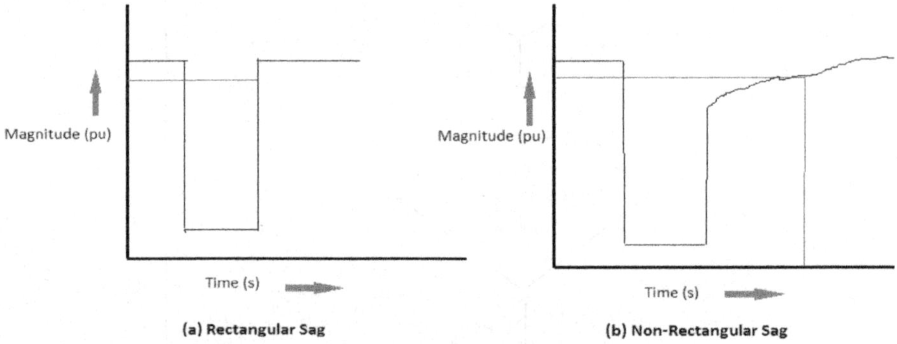

Figure 3.3: Rectangular and Non-rectangular sag

To overcome from this problem a new method is applied in which voltage during sag also include values of average voltage, sag voltage, minimum voltage and durations [7]. In this new method both type of sag rectangular and non-rectangular are calculated and each equipment's ride-through capability can also be calculated out. So, during calculating AV both rectangular and non-rectangular sag must be taken into account. According to equipment's sensitivity to VS three classification are mentioned by power engineers are as follows -

1. Equipments those which are sensitive to magnitude of VS.
2. Equipments those which are sensitive to both magnitude and duration of VS.
3. Equipments which are sensitive to characteristics of VS other than magnitude and duration.

3.4 Overview of IEEE towards Voltage Sag

According to IEEE voltage sag is decrease in RMS value of voltage for duration of 0.5 cycle to 1 minute at typical value of 0.1 to 0.9 pu. They recommended that to obtaine numerical value of VS take " a sag to 20% means line voltage is decreased to 20% of the normal, not reduced by 20%". As per IEEE, to perform compatible evaluation between system and equipment's VS data is very important. Utilities for specific areas may not have enough sag information, they may have information regarding actual measurements, predicted performance. Exact complete measured information for the facility obtained over many years of observation as long as the system remains the same in future also. Even utility data due to year, season, location, level of voltage and other factors may be varied. Since, distribution circuits are changed or modified as per need frequently. Where VS is being measured is an important point. Because utility data may be taken from utility connection to facility, nearby sites or at substation. In this case equipments are installed inside the substation, where effect of VS may not be that much similar to those of facility side. So, it is the good option to measure the sag and take data as close to the feeder connected to equipment. In case of actual data is not present, apply judgment at the place of data. To collect the data it is suggested to install monitors and collect data regarding sag. The longer period of time for which data is collected the more reliable data will be obtained. Whether is one of the most important reason which affect the frequency of sag. In IEEE Gold book gives details of techniques and guidelines for VS.

Most of the VS generated on utility system, the voltage operating range normally can create difference between ride-through and disruption. The nominal rating of voltage is higher compared to those of equipments rating, this is for the reason to allow drop in voltage within the facility electrical distribution network. The design should permit 5% of voltage drop. Excessive of drop in voltage will make the equipment to operate at remaining voltage without concerning its sag capability. And also lower voltage can cause increase in temperature, reduced speed and torque etc. Tapping of transformer gives an inexpensive way to compensate voltage drop. Winding configuration of transformer affects the magnitude of sag. For SLG fault on utility, the magnitude of sag depends on the sequence and number of delta-wye transformation. Zero-volts on one of the phase incoming becomes 33% on the secondary side of transformer (58% for loads which are neutral connected). Three phase transformers and configuration will create different internal sag for same incoming sag. In IEEE standard 1346-1998 it is concluded that sag contour lines work appreciably good with rectangular sensitivity curve whereas, non-rectangular sensitivity curve will require a little more effort [10].

According to IEEE standard 1159-2009 electronic devices those which have proper battery backup should remain unaffected by VS [3]. Equipments like transformer cables, switchgears, bus, CTs and PTs should not come into damage and malfunctioning due to VS. A little bit change in induction machine and a small reduction in output from capacitor can occur at the time of sag. It is concluded that the PQ monitor record the event as sag but the equipments take it as interruption. Sag is a product of drop in voltage of line and neutral conductors, while neutral to ground voltage consider only neutral.

*****

CHAPTER 4
RMS METHOD

4.1 Why RMS Method is Used?

Generally, RMS method is used for periodic signals [14]. Here, it is used as to give information of periodic signal which becomes non- periodic due to PQD or VS. From the pure periodic signal only those places where the signal is not periodic can be calculated out easily with the help of RMS method. RMS stand for root mean square value of voltage. Two types of RMS are continuous and discrete RMS. Discrete RMS voltage contains two voltages or say samples one before the event occurred and another after the event, which in turn gives the RMS value between the value of previous and new RMS voltage. In continuous, RMS values are updated and a new sample is obtained every time. At the time of transition, RMS calculated does not give the exact voltage value of new state, until the window entirely contains the samples of new state. RMS is the best qualitative quantity which is proved to be very helpful in fetching out the details of the whole power system. RMS method is very helpful to describe the percentage voltage sag in the system.

To classify RMS variations magnitude and duration of disturbances are taken as the basis. Characterization of RMS variations are structured in three levels which are as follows [15]-

1. Phase or Component event level-

Each phase RMS variation measurement may contain number of components. Those RMS variations which have rectangular shape are characterize by single magnitude and duration but those variations which are non-rectangular and have number of components can not be characterize by single magnitude and duration.

2. Measurement event level-

Fault can effect one, two or all three line, magnitude and duration of final RMS variation for three phases may be substantially different. In this case magnitude and duration of that phase which have highest deviation of voltage from nominal voltage is taken.

3. Aggregate event level-

It is collection of measurement linked with a single system occurence in a single set of event characteristics.

4.2 RMS Calculation

To calculate the voltage sag the lowest remaining RMS voltage is calculated out. The voltage remained last in all the three phases among which the lowest voltage is consider.

N= Number of samples per period

then the RMS voltage at point "a" is

$$V_{RMS}(b)=\sqrt{\left[\frac{1}{N}\sum_{i=b-N+1}^{i=b}(v_i)^2\right]} \qquad 4.1$$

VS written in function form as

$$V(t)=\begin{cases} V_b \sin(wt) & t < t_1 \\ V_b \sin(wt) & t > t_2 \\ V_{b\,sag} \sin(wt+\Phi) & t_1 < t < t_2 \end{cases} \qquad 4.2$$

Where,

w = Angular frequency

V_b= Peak before fault voltage

$V_{b\,sag}$= Peak voltage at the time of sag

Φ = Phase angle shift

t_1= Time of initiation of sag

t_2= Time of recovering of sag

$\Delta t = t_2 - t_1$ = Sag Duration

4.3 Six Phase RMS Method

In this method zero-sequence component is removed. As zero-sequence voltage does not affect the sag type. So, we get

$$V_0 = \frac{1}{3}(v_a + v_b + v_c) \quad (4.3)$$

So, the RMS voltage for the three phase voltage are

$$V_A = \text{RMS}\{v_a - \tfrac{1}{3}(v_a + v_b + v_c)\} \quad (4.4)$$

$$V_B = \text{RMS}\{v_b - \tfrac{1}{3}(v_a + v_b + v_c)\} \quad (4.5)$$

$$V_C = \text{RMS}\{v_c - \tfrac{1}{3}(v_a + v_b + v_c)\} \quad (4.6)$$

RMS voltage for three phase to phase voltages are

$$V_{AB} = \text{RMS}\,\frac{(v_a - v_b)}{\sqrt{3}} \quad (4.7)$$

$$V_{BC} = \text{RMS}\,\frac{(v_b - v_c)}{\sqrt{3}} \quad (4.8)$$

$$V_{CA} = \text{RMS}\,\frac{(v_c - v_a)}{\sqrt{3}} \quad (4.9)$$

According to this method sag type can be determine as shown in table.

Table 4.1: Sag type for minimum RMS voltage on different phases

Minimum RMS voltage	Sag type
V_A	D_a
V_B	D_b
V_C	D_c
V_{AB}	C_c
V_{BC}	C_a
V_{CA}	C_b

Where,

D_a is drop in phase A

D_b is drop in phase B

D_b is drop in phase B

C_a is drop in between phase B and C

C_b is drop in between phase A and C

C_c is drop in between phase A and B

4.4 Voltage Sag Indices

Voltage sag indices is divided in three parts namely voltage sag energy index, voltage sag severity index, SARFI-X. To calculate VS in the power system, VSI are very important. VSI gives the detail of VS mathematically. In VSI a database is prepare to analyze the VS event.

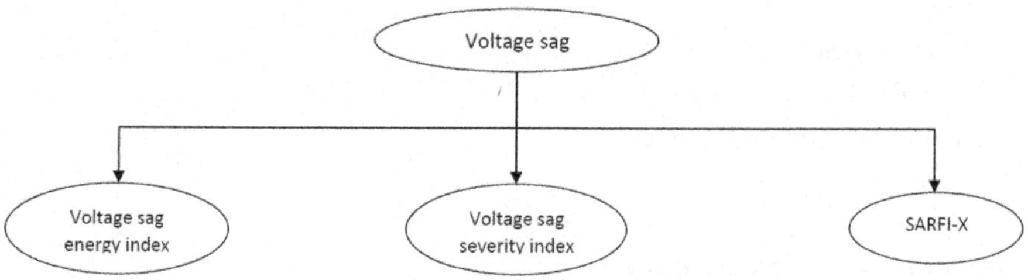

Figure 4.1: Parts of Voltage Sag Indices

4.4.1 Voltage Sag Energy Index

VS energy index is one of the part of VS indices it is calculated as

$$E_{VS}= \int_0^T \left\{ 1 - \left[\frac{(V_{RMS})_{DE}}{V_{nominal}}\right]^2 \right\} dt \tag{4.10}$$

Where,

$V_{(RMS)\,DE}$ = RMS voltage during event

$V_{nominal}$ = Nominal voltage

Unit of E_{VS} that is VS energy index is time. Here, integration is taken in order to calculate all the RMS voltage values which are below the threshold limit during the period of event.

By formula (4.1) for half-cycle

$$V_{RMS}(b)= \sqrt{\left[\frac{1}{N}\sum_{i=b-N+1}^{i=b} (v_i)^2\right]} \tag{4.11}$$

So, VS energy index for half cycle

$$E_{VS}= \frac{1}{2f} \sum_{b=1}^{N} \left\{ 1 - \left[\frac{\left(V_{RMS\frac{1}{2}}\right)_{DE}}{V_{nominal}}\right]^2 \right\} dt \tag{4.12}$$

Where,

f = Frequency

$\sum$ is used to calculate all the values of RMS voltage during the VS event.

If only residual or remaining voltage and duration of an event is known then the RMS value of voltage is assumed as constant.

which give E_{VS} as

$$E_{VS}= \left\{ 1 - \left[\frac{(V)_{res}}{V_{nominal}}\right]^2 \right\} \times T \tag{4.13}$$

Where,

T= Duration

V_{res}= Residual voltage

In case of three phase systems

$$E_{VS} = E_{VS_A} + E_{VS_B} + E_{VS_C} \quad (4.14)$$

4.4.2 Voltage Sag Severity

Voltage sag severity is one of the another most important part of the VSI. It is also calculated by using residual voltage and duration of VS in combination with reference curve. It is denoted by S.

$$S = \left[\frac{1 - V_{res}}{1 - V_{curve}(T)}\right] \quad (4.15)$$

V_{res}= Residual voltage

V_{curve} (T) = Residual voltage of the reference curve which is of same duration.

4.4.3 SARFI-X

SARFI stands for system average RMS (variation) frequency index and X denotes voltage. It gives the average number of mentioned RMS variation measurement events that occurred for a particular duration of time for magnitude less than X. Thus SARFI gives average voltage quality of the system.

Mathematically, SARFI-X is given by

$$\text{SARFI-X} = \frac{\sum N_E}{D} \quad (4.16)$$

Where,

X= RMS voltage threshold

N_E= Number of events

D= Number of days measuring these events

As SARFI-X is also used for calculating frequency of occurrence of varying RMS of various magnitude. There are three indices of SARFI-X which are further made as per IEEE standards 1159 which are named as [22]

1. System Instantaneous Average RMS (variation) Frequency Index (SIARFI-X)

$$\text{SIARFI-X} = \frac{\sum NI_i}{D} \quad (4.17)$$

NI_i = Number of customers experiencing instantaneous voltage deviations.

2. System Momentary Average RMS (variation) Frequency Index (SMARFI-X)

$$\text{SMARFI-X}=\frac{\sum NM_i}{D} \tag{4.18}$$

NM_i = Number of customer experiencing momentary voltage deviations.

3. System Temporary Average RMS (variation) Frequency Index (STARFI-X)

$$\text{STARFI-X}=\frac{\sum NT_i}{D} \tag{4.19}$$

NT_i = Number of customer experiencing temporary voltage deviations.

SARFI-curves is one of the main part SARFI-X. They give the details of rate of voltage sag below an equipment's compatibility curve. SARFI-curve also gives another three types of indices on the basis of threshold value. These are as follows-

$SARFI_{CBEMA}$

$SARFI_{ITIC}$

$SARFI_{SEMI}$

$SARFI_{CBEMA}$ curve is valid for events whose duration is greater than 1/2 cycle. They does not limit the duration of RMS variation to 60 seconds [24].

$SARFI_{ITIC}$ curve is the new CBEMA curve [23]. This curve was developed by Information Technology Industry Council in order to get exact performance of electronic equipments. This curve give information about AC input which can be tolerated by most of the electronics equipments. In this curve there are two curves one represent upper limit and another one represent lower limits. Events above the upper curve or below the lower curve are taken as the cause of malfunctioning of equipments.

Table 4.2: Method for calculating SARFI-ITIC

Residual voltage by ITIC curve	Duration
0.0 pu	1 ms < D ≤ 20 ms
< 0.7 pu	20 ms < D ≤ 500 ms
<0.8 pu	500 ms < D < 20 ms
<0.9 pu	10 s < D

$SARFI_{SEMI}$ curve gives the details for VS when voltage is below the SEMI curve which can cause damage of semiconductor manufacturing equipment.

Table 4.3: Method for calculating SARFI-SEMI

Residual voltage by SEMI curve	Duration
0.0 pu	D ≤ 20 ms
< 0.5 pu	20 ms < D ≤ 200 ms
< 0.7 pu	200 ms < D ≤ 500 ms
< 0.8 pu	500 ms < D ≤ 10 ms
< 0.9 pu	10 s < D

*****

CHAPTER 5
FAULTS

5.1 Single Line To Ground Fault (SLG)

In SLG, fault take place between the line and the ground. About 65-70 % of faults are SLG fault. It occurs due to short circuit fault, this short circuit arises in between any one phase with the ground. It may be occur due to breaking of line and come in contact to the ground or may be due to insulation failure. VS due to this type of faults give birth to suppressed voltage with high phase angle jump in faulted line. While in other two lines phase angle remain undisturbed. To simulate this type of fault all the three sequence network are used and also current in all these sequences are same in phase angle and magnitude [16]. So, all the sequence network are connected in series.

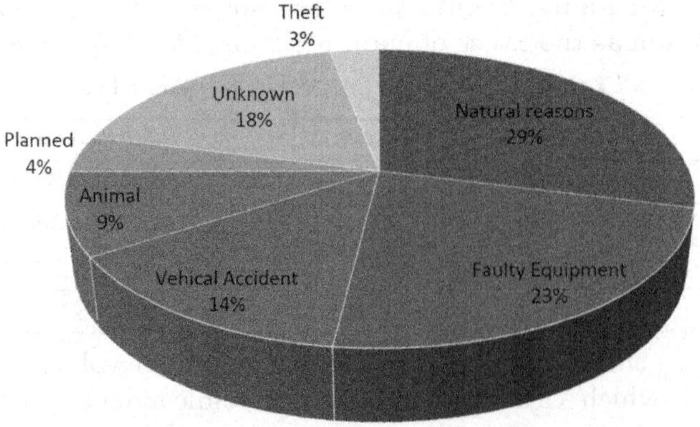

Figure 5.1: Percentage of different types of fault

5.2 Double Line (DL) and Double Line to Ground Fault (DLG)

In DL fault, zero-sequence component of current is not present and relation of positive sequence with negative sequence is such that both have component of current those which are equal in magnitude but opposite in phase. To simulate this fault there is no need of zero sequence but the positive sequence and the negative sequence are connected in opposition. Also, in this positive sequence component of voltage is equal to the negative sequence component of voltage. About 5-10 % of faults are DL fault.

In DLG, all the three that is zero, positive and negative sequence network are required to simulate this type of fault. In this negative and zero sequence network are connected in parallel first then in opposite with positive sequence network. Around 15-20 % of faults are DLG fault.

Table 5.1: Fault probability in different type of power system component

Power System Component	Fault Probability (in %)
Overhead lines	50
Switchgears	12
CT,PT, relays, control equipment etc	12
Transformers	10
Underground cables	9
Generators	7

5.3 Three Phase Fault (3-Φ)

In 3-Φ fault as the name suggest fault occur in all the three phases or lines. In this type of fault short circuit occurred in between all the three phases. In this zero and negative sequence current component are absent, only positive sequence current is only present. Positive sequence current is not affected by mutual coupling and is equal to phase current.

Table 5.2: Percentage of different fault initiators

Fault Initiators	% of all faults
Normal deterioration from age	7
Exposures to chemicals	9
Other mechanical damage	12
Exposure to dust	14.5
Shorting by tools, rodents etc	18
Exposure to moisture	22.5

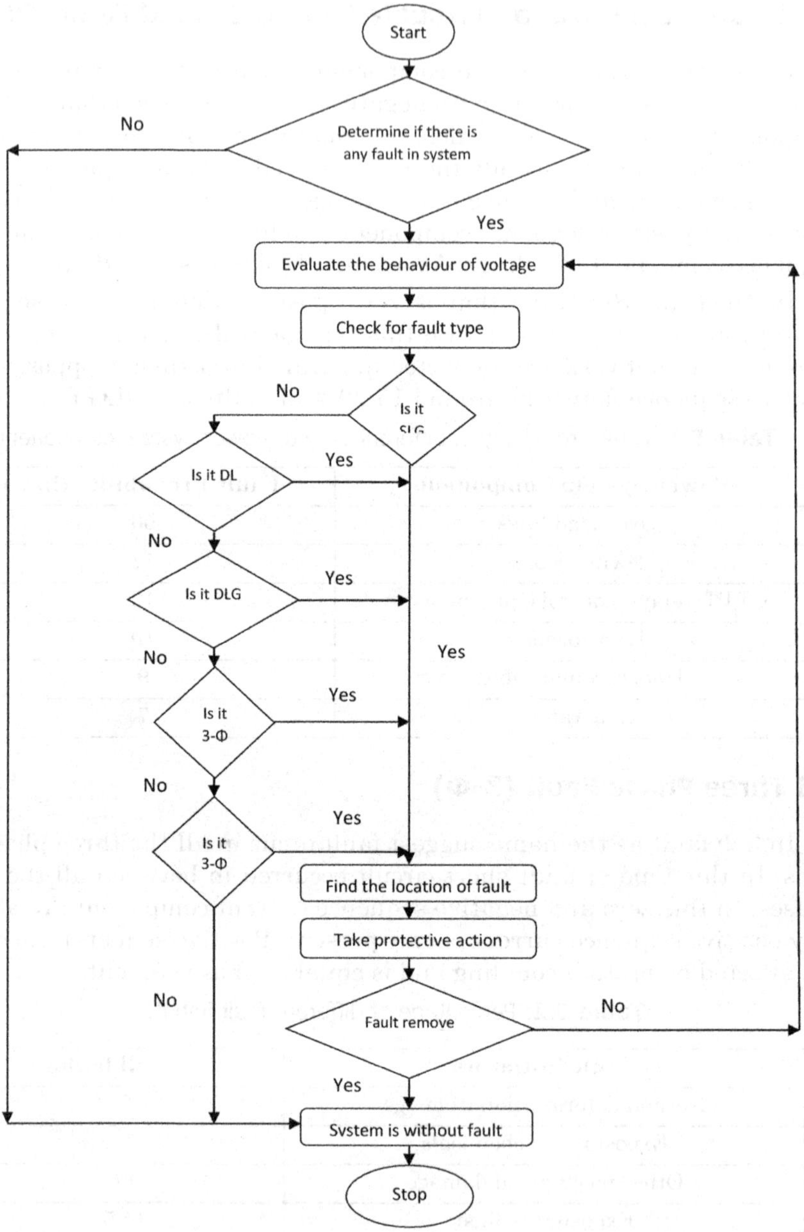

Figure 5.2: Flowchart for fault analysis

*****

CHAPTER 6
COMPARISION WITH SIMULATION

In this dissertation work, VS on 33/11 kv line is measured. For this Madhya Pradesh Power Transmission Company feeding Shakti Nagar is taken as a case study. In this, VS by applying different type of faults on different types of transformer connection has been studied. By this percentage sag on different transformer connection during various fault condition has been calculated. With the help of this, one can study the behaviour or say transformer connection affects on VS. Transformer energizing severely causes VS. Transformer energizing time and also other characteristics straightly linked or in other words depends on transformer connections. So, here by studying this affect of transformer connection on different types of faults regarding VS graphs has been obtain from MATLAB/SIMULINK. Figure 6.1 shows Single Line diagram of the feeder for Shakti Nagar consider under study.

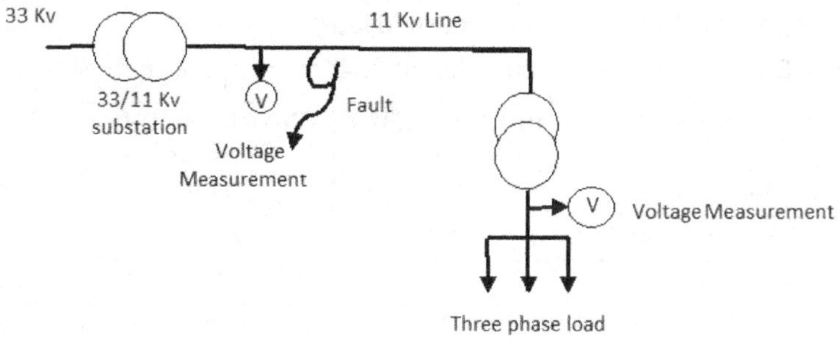

Figure 6.1: Single line diagram

Table shown below gives the details of parametric configuration taken under MATLAB/ SIMULATION.

Table 6.1: Parameters of MATLAB/SIMULINK model

COMPONENT	PARAMETER
Source	33/11Kv, X/R= 10
Transformer	200KVA, 11/0.4 Kv
Frequency	50 Hz
Load	140 KW, 10 KVAR

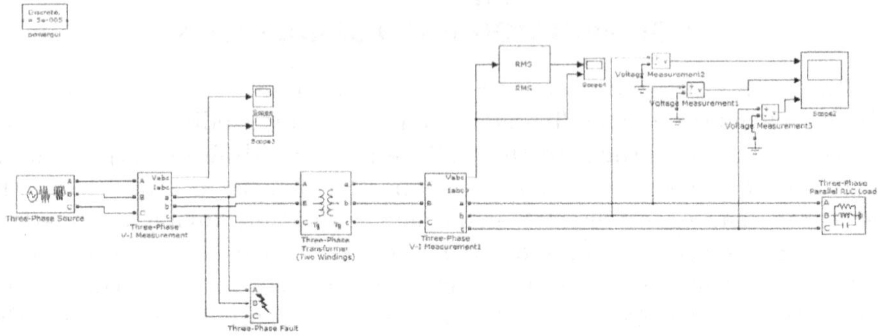

Figure 6.2: MATLAB model

6.1 Transformer Y_g-Y_g Connected

For Y_g-Y_g transformer connection different types of fault giving birth to VS is shown in table below with percentage sag.

Table 6.2: Percentage sag on different fault on Y_g-Y_g transformer connection

Type of fault	Pre-fault RMS Voltage	Post-fault RMS Voltage	% Sag
Single line to ground	220	34.97	84.10
Double line to ground	220	167.32	23.94
Double line	220	196.16	10.83
Three phase to ground	220	172.48	21.60
Three phase	220	213.24	3.07

(a)

(b)

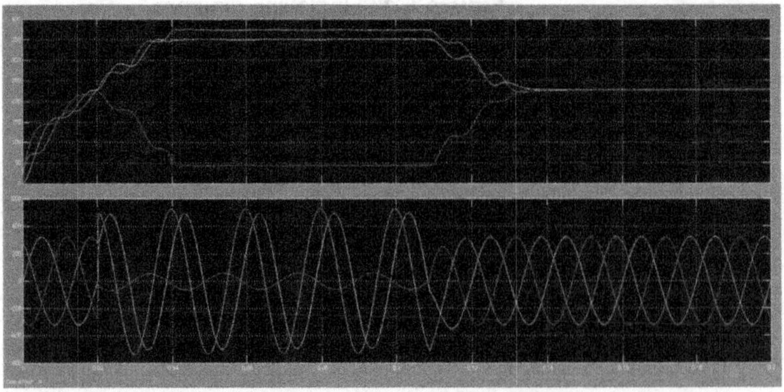

Figure 6.3: SLG fault (a) RMS waveform (b) Instantaneous waveform

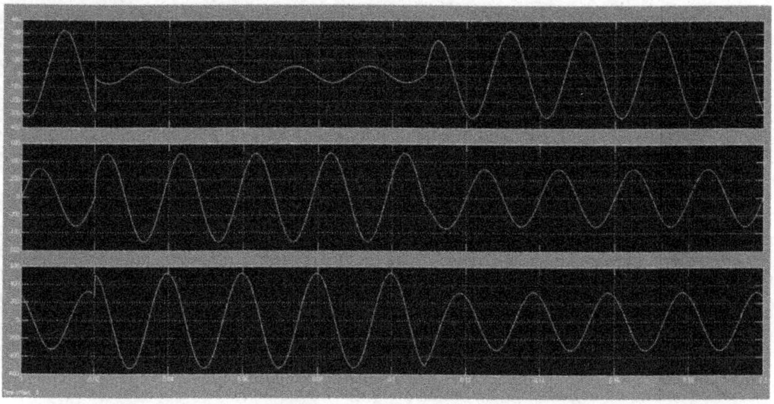

Figure 6.4: Voltage waveform of all three phases during SLG

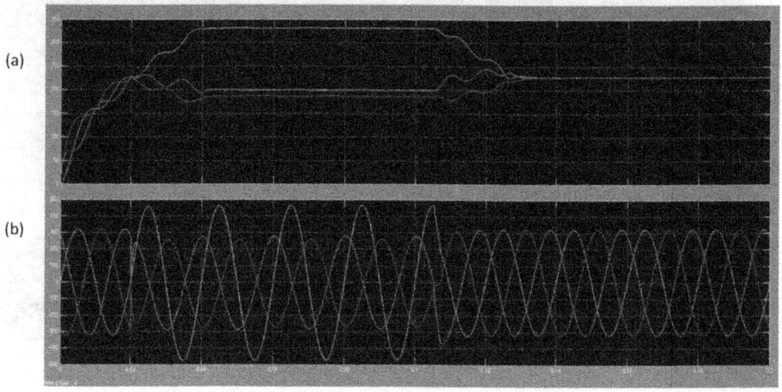

Figure 6.5: DLG fault (a) RMS waveform (b) Instantaneous waveform

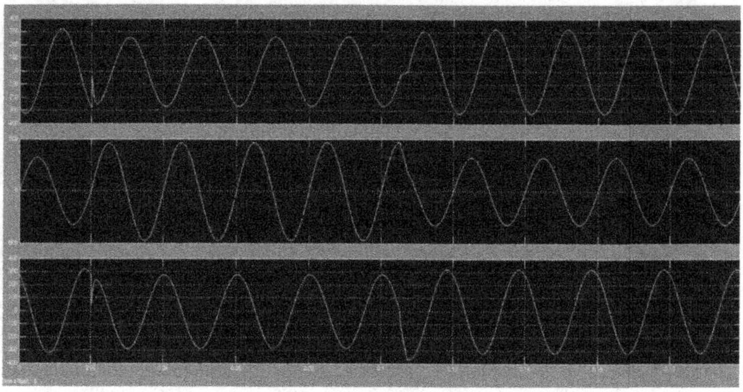

Figure 6.6: Voltage waveform of all three phases during DLG

(a)

(b)

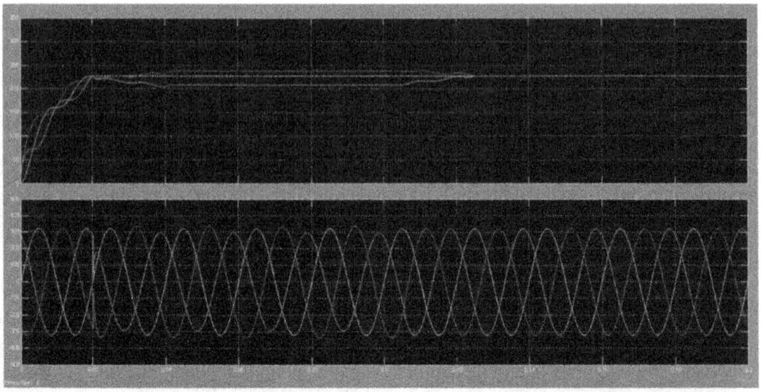

Figure 6.7: DL fault (a) RMS waveform (b) Instantaneous waveform

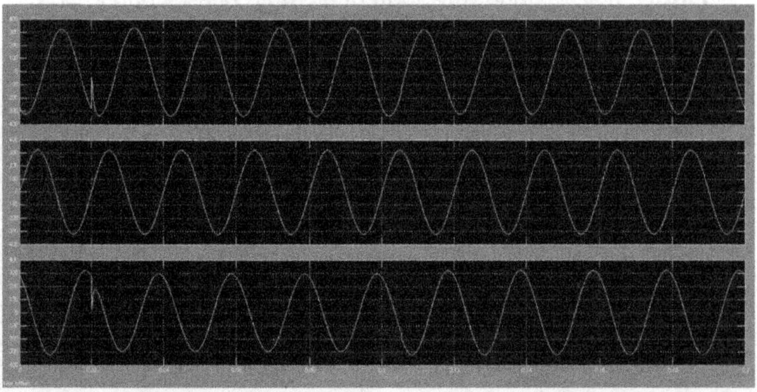

Figure 6.8: Voltage waveform of all three phases during DL

(a)

(b)

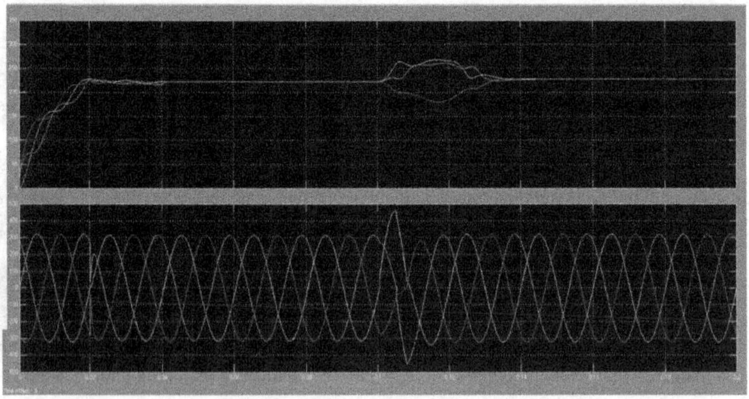

Figure 6.9: 3-Φ G fault (a) RMS waveform (b) Instantaneous waveform

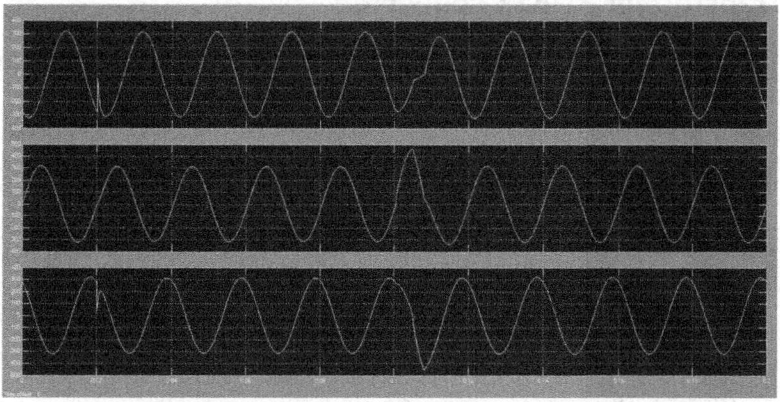

Figure 6.10: Voltage waveform of all three phases during 3-Φ G

(a)

(b)

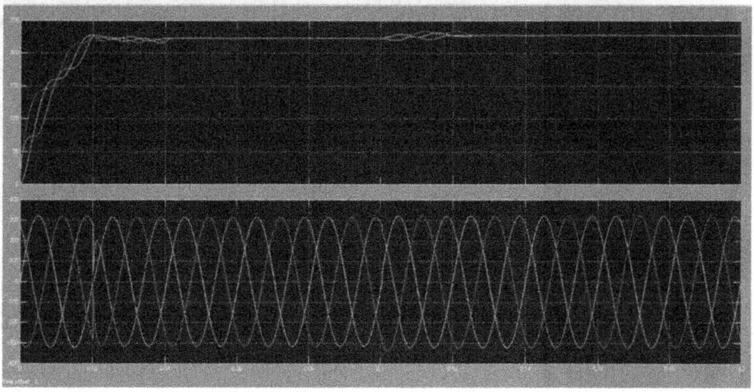

Figure 6.11: 3-Φ fault (a) RMS waveform (b) Instantaneous waveform

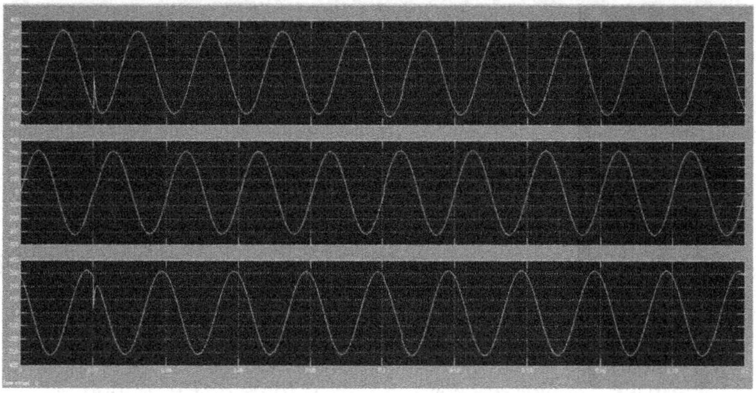

Figure 6.12: Voltage waveform of all three phases during 3-Φ

6.2 Transformer Y_g-Δ11 connected

For Y_g-Δ11 transformer connection different types of fault giving birth to VS is shown in table below with percentage sag.

Table 6.3: Percentage sag on different fault on Y_g-Δ11 transformer connection

Type of fault	Pre-fault RMS Voltage	Post-fault RMS Voltage	% Sag
Single line to ground	220	202.90	7.77
Double line to ground	220	204.90	6.86
Double line	220	202.88	7.78
Three phase to ground	220	214.79	2.36
Three phase	220	214.04	2.70

(a)

(b)

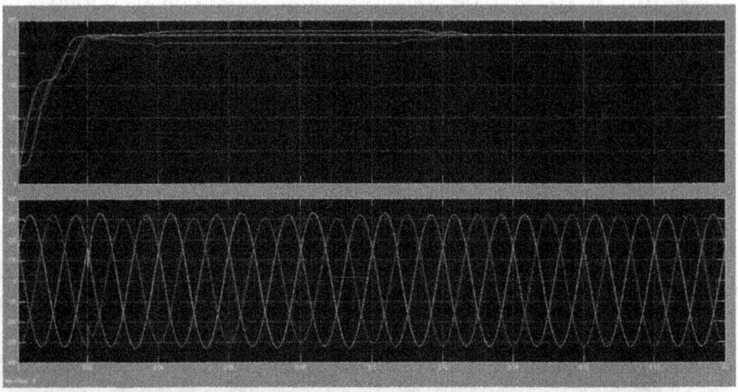

Figure 6.13: SLG fault (a) RMS waveform (b) Instantaneous waveform

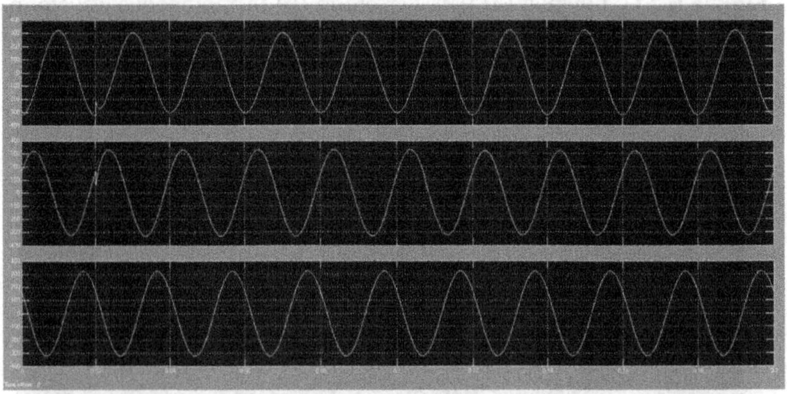

Figure 6.14: Voltage waveform of all three phases during SLG

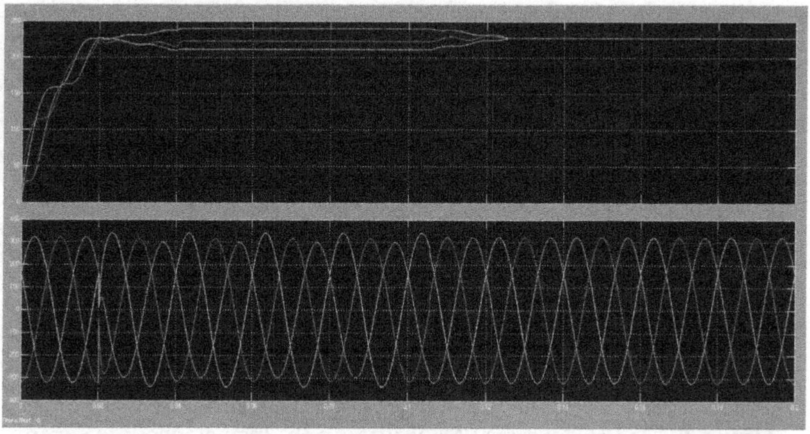

Figure 6.15: DL fault (a) RMS waveform (b) Instantaneous waveform

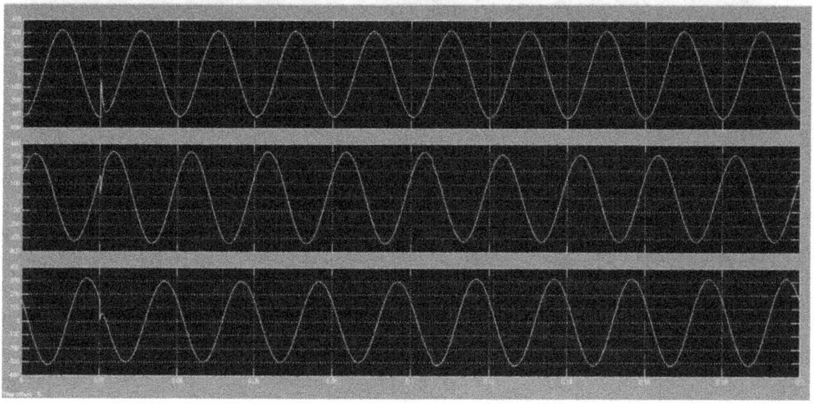

Figure 6.16: Voltage waveform of all three phases during DL

6.3 Transformer Y_g-Y connected

For Y_g-Y transformer connection different types of fault giving birth to VS is shown in table below with percentage sag.

Table 6.4: Percentage sag on different fault on Y_g-Y transformer connection

Type of fault	Pre-fault RMS Voltage	Post-fault RMS Voltage	% Sag
Single line to ground	220	216.52	1.58
Double line to ground	220	196.35	10.75
Double line	220	196.16	10.83
Three phase to ground	220	213.32	3.03
Three phase	220	213.24	3.07

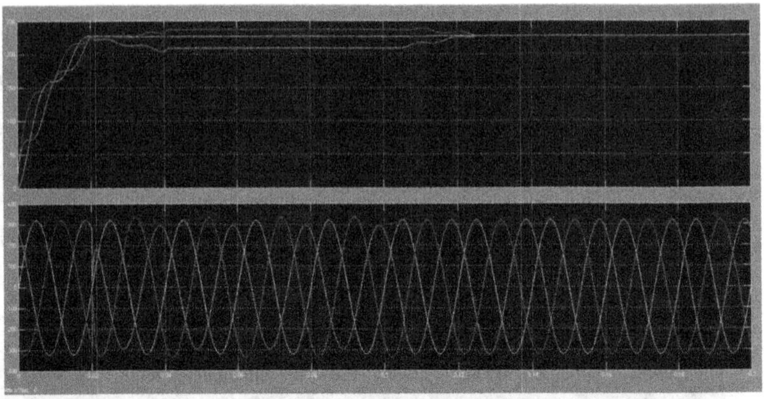

Figure 6.17: DLG fault (a) RMS waveform (b) Instantaneous waveform

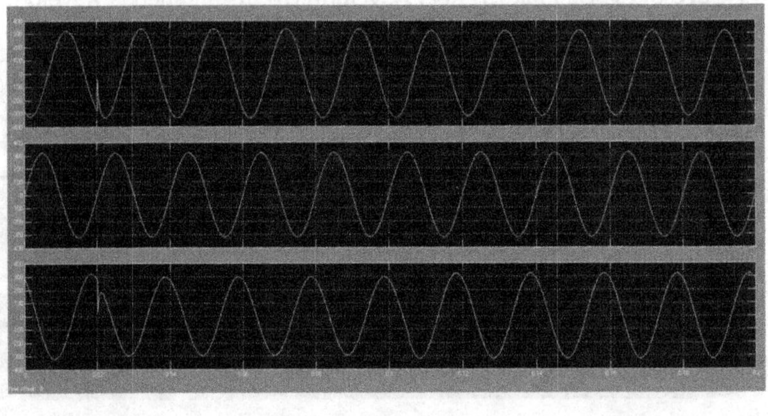

Figure 6.18: Voltage waveform of all three phases during DLG

*****

CHAPTER 7
CONCLUSION AND FUTURE SCOPE

7.1 Conclusion

From this thesis work it is concluded that during SLG fault when transformer is Y_g-Y_g connected VS is quite high compare to transformer when Y_g-Y connected. Similarly, in case of 3-Φ fault VS is observable when transformer is Y_g-Y_g connected. When transformer connection is Y_g-Y DL fault is observable more clearly then in transformer connection Y_g-Δ11.

7.2 Future Scope

VS can be simulated in many other software such as MiPower but here it is simulated on MATLAB. With the advancements in software tools we get even more precise and exact result. Here, simulation perform by applying various types of faults on different types of transformer connection. In future it can be used as to design mitigation or say reduction methodology at a cheap cost other then using FACTS devices to reduce VS by transformer connection design and reducing the requirement of compensating devices to the appreciable extent or limit. In this way power engineers must work, not only to reduce PQD but also to design economical power system which is quite efficient and at the same time must have the ability of reducing VS by less use of compensating devices or say without using unnecessary compensating devices where it is possible.

REFERENCES

[1] Joaquín Caicedo, Felipe Navarro, Edwin Rivas and Francisco Santamaría Electromagnetic Compatibility and Interference Group (GCEM) (2012) Voltage Sag Characterization with Matlab/Simulink, Colombia IEEE

[2] S.Kamble and Chandrashekhar Thorat (2014) Voltage Sag Characterization in a Distribution Systems : A Case Study Journal of Power and Energy Engineering, 2014,2,546-553.

[3] (2009) IEEE Recommended Practice for Monitoring Electric Power Quality. IEEE Std. 1159-2009.

[4] P.Heine and M. Lehtonen, (2003) Voltage Sag Distributions Caused by Power Systems Faults. IEEE Transactions on Power Systems, **18**, 1367-1373. http://dx.doi.org/10.1109/TPWRS.2003.818606

[5] R. Naidoo and P.Pillay (2007) A New Method of Voltage Sag and Swell Detection. IEEE Transactions on Power Delivery, **22**, 1056-1063. http://dx.doi.org/10.1109/TPWRD.2007.893185

[6] E.E. Juarez and A. Hernandez (2006) An Analytical Approach for Stochastic Assessment of Balanced and Unbalanced Voltage Sags in Large Systems. IEEE Transactions on Power Delivery, **21**, 1493-1500. http://dx.doi.org/10.1109/TPWRD.2005.860266

[7] D.-J. Won, S.-J. Ahn and Y.Chung,(2003) A New Definition of Voltage Sag Duration Considering the Voltage Tolerance Curve. IEEE Bologna Power Tech Conference, Bologna, **4**, 23-26 June 2003.

[8] L.D. Zhang and M.H.J.Bollen,(2000) Characteristic of Voltage Dips (Sags) in Power Systems. IEEE Transactions on Power Delivery, **25**, 827-832.

[9] Larry Conard, Kevin Little and Cliff Grigg (1991) Predicting And Preventing Problems Associated With Remote Fault-Clearing Voltage Dips. IEEE Transactions on Industry Applications,27.

[10] (1998) IEEE Recommended Practice for Evaluating Electrical Power System Compatibility With Electronic Process Equipment.IEEE Std.1346-1998.

[11] F.R.Zaro, M.A.Abibo, S.Ameenuddin and I.M.Elamin (2012) Characterization Of Short-Duration Voltage Events. IEEE International Conference on Power and Energy(PECon).

[12] S.Khokhar, A.A. Mohd Zin, A.S.Mokhtar, NAM Ismail (2014) MATLAB/ Simulink Based Modelling and Simulation of Power Quality Disturbances. IEEE.

[13] Ohrstrom, M. and Soder, L. (2003) A Comparison of Two Methods Used for Voltage Dip Characterization. IEEE Power Tech Conference, Bologna, 4, 23-26 June 2003.

[14] Styvaktakis, E., Bollen, M.H.J. and Gu, I.Y.H. (2002) Automatic Classification on Power System Event Using RMS Voltage Measurements. IEEE Power Engineering Society Summer Meeting, 2, 824-829.

[15] Electrical Power System Quality by Roger C. Dugan, Mark F. McGranaghan, Surya Santoso, H.Wayne Beaty.

[16] Electrical Power Systems (6 th edition) by C.L.Wadhwa.

[17] Power System Analysis by Prabha Kundur

[18] Jorge Esteves, Karstein Brekke, Kevin Niall, Maurizio Delfanti, Math Bollen (2011) Voltage Quality Monitoring, Dip Classification And Responsibility Sharing.

[19] Manish N. Sinha, Dr. B.R. Parekh (2014) Simulation Of Voltage Sag Magnitude Estimation In A Power System Network, (IJAREEIE) International journal of advanced research in electrical electronics and instrumentation engineering.

[20] M.H.J. Bollen, L.D. Zhang, A Method For Characterization Of Three-Phase Unbalanced Dips From Recorded Voltage Waveshapes.

[21] Noridayu Bte Zainal (2007) Voltage Sag In Distribution System Using PSCAD Software.

[22] Roberto Chouhy Leborgne (2005) Voltage Sag Characterization And Estimation.

[23] Frantisek Kinces (2004) Voltage Sag Indices And Statistics.

[24] Alexis Polycarpou (2011) Power Quality And Voltage Sag Indices In Electrical Power Systems.

EPILOGUE

The energy sector is a critical aspect of modern society, and power system simulation software tools are indispensable in ensuring the reliability and stability of the grid. ETAP, MiPOWER, CYME, and Simulink are among the most widely used power system simulation software tools. Each of these tools has its unique features, advantages, and disadvantages, making them suitable for different applications. Here we compare ETAP, MiPOWER, CYME, and Simulink in terms of their features, capabilities, ease of use, and suitability for different applications.

ETAP

ETAP is a comprehensive power system analysis and design software tool that supports the analysis of power systems ranging from simple to complex. ETAP is an acronym for Electrical Transient Analyzer Program, and it is developed by Operation Technology, Inc. (OTI) based in Irvine, California. ETAP is widely used by power engineers, consultants, and researchers worldwide due to its comprehensive analysis capabilities, accuracy, and versatility.

ETAP offers a wide range of modules and capabilities for power system analysis, including load flow analysis, short circuit analysis, transient stability analysis, harmonic analysis, motor acceleration analysis, and relay coordination analysis. ETAP also supports system optimization, and it offers a range of tools for power system design, including cable sizing, transformer sizing, and capacitor placement.

ETAP is designed to be user-friendly, and it offers a graphical user interface that allows users to model power systems using drag and drop functionality. ETAP also supports data exchange with other software tools, including AutoCAD and Microsoft Excel.

MiPOWER

MiPOWER is a power system simulation software tool that offers comprehensive analysis and design capabilities for power systems.

MiPOWER is developed by PowerWorld Corporation based in Champaign, Illinois, and it is widely used by power engineers, consultants, and researchers worldwide.

MiPOWER offers a wide range of modules and capabilities for power system analysis, including load flow analysis, short circuit analysis, transient stability analysis, harmonic analysis, and contingency analysis. MiPOWER also supports system optimization, and it offers a range of tools for power system design, including cable sizing, transformer sizing, and capacitor placement.

MiPOWER is designed to be user-friendly, and it offers a graphical user interface that allows users to model power systems using drag and drop functionality. MiPOWER also supports data exchange with other software tools, including AutoCAD and Microsoft Excel.

CYME

CYME is a power system simulation software tool that offers comprehensive analysis and design capabilities for power systems. CYME is developed by Eaton Corporation based in Pittsburgh, Pennsylvania, and it is widely used by power engineers, consultants, and researchers worldwide.

CYME offers a wide range of modules and capabilities for power system analysis, including load flow analysis, short circuit analysis, transient stability analysis, harmonic analysis, and contingency analysis. CYME also supports system optimization, and it offers a range of tools for power system design, including cable sizing, transformer sizing, and capacitor placement.

CYME is designed to be user-friendly, and it offers a graphical user interface that allows users to model power systems using drag and drop functionality. CYME also supports data exchange with other software tools, including AutoCAD and Microsoft Excel.

Simulink

Simulink is a power system simulation software tool that is part of MATLAB, a widely used numerical computing environment developed by MathWorks based in Natick, Massachusetts. Simulink is widely used by power engineers, consultants, and researchers worldwide.

Simulink offers a wide range of modules and capabilities for power system analysis, including load flow analysis, short circuit analysis, transient stability analysis, and harmonic analysis. Simulink also supports system optimization, and it offers a range of tools for power system design, including cable sizing, transformer sizing, and capacitor placement.

While Simulink is a powerful and versatile tool for power system simulation, there are some limitations to its capabilities. Here are some of the most notable limitations of Simulink:

1. Steep Learning Curve: Simulink can be difficult to learn for new users, especially those who are not familiar with the MATLAB environment. This is because Simulink requires a strong understanding of programming concepts and the MATLAB language.
2. Limited Analysis Capabilities: Compared to other power system simulation software tools like ETAP and CYME, Simulink's analysis capabilities are more limited. Simulink does not support certain types of analysis, such as relay coordination analysis or load flow analysis with unbalanced loads.
3. Limited Library of Components: While Simulink has a library of pre-built components for power system simulation, the library is not as extensive as those of other software tools like ETAP and CYME. This means that users may need to create their own custom components for their simulations.
4. Performance Issues: Simulink can be slow and resource-intensive when simulating large and complex power systems. This can result in longer simulation times and slower performance.
5. Lack of Real-Time Simulation Capabilities: Simulink is not well-suited for real-time power system simulation applications. This is because Simulink's simulations are typically run in a time-stepped manner, which is not suitable for real-time applications that require continuous simulation.

Overall, while Simulink is a powerful tool for power system simulation, its limitations mean that it may not be the best choice for all applications. Users should carefully evaluate their simulation needs and consider other software tools like ETAP and CYME if they require more advanced analysis capabilities or real-time simulation.

Made in the USA
Monee, IL
04 November 2026

67381188R00115